International Conference on Historic Ships 2020

Online
2 - 3 December 2020

ISBN: 978-1-7138-8299-2

Printed from e-media with permission by:

Curran Associates, Inc.
57 Morehouse Lane
Red Hook, NY 12571

Some format issues inherent in the e-media version may also appear in this print version.

The Royal Institution of Naval Architects
8-9 Northumberland Street
London, WC2N 5DA
United Kingdom

Phone: 020 7235 4622
Fax: 020 7259 5912

publications@rina.org.uk

Additional copies of this publication are available from:

Curran Associates, Inc.
57 Morehouse Lane
Red Hook, NY 12571 USA
Phone: 845-758-0400
Fax: 845-758-2633
Email: curran@proceedings.com
Web: www.proceedings.com

International Conference

Historic Ships 2020

2-3 December 2020, Online Conference

PROCEEDINGS

THE ROYAL INSTITUTION OF NAVAL ARCHITECTS
8-9 Northumberland Street, London, WC2N 5DA, Telephone: +44 (0)20 7235 4622

The Royal Institution of
Naval Architects

INTERNATIONAL CONFERENCE

Historic Ships 2020

2-3 December 2020, Online Conference

Papers published in these conference Proceedings are peer reviewed.

THE ROYAL INSTITUTION OF NAVAL ARCHITECTS
8-9 Northumberland Street
London WC2N 5DA

Telephone: 020 7235 4622
Fax: 020 7259 5912

CONTENTS

A MYSTERY OF THE FIRST ICEBREAKER

D. Ponkratov, J. Pérez-Martínez, Royal Institution of Naval Architects, UK

SUMMARY

Almost a detective story describing a hopeless search in archives which took many years and apparently resulted in the discovery of not just the singular origin of the first ice breaker, but the engineering ingenuity that led to that first design. This paper will uncover the forgotten history of the first icebreaker and will recreate its original shape using the original general arrangements and references uncovered during the research.

1. INTRODUCTION

Some innovations live relatively short lives - they take off, make contribution and disappear forever (who builds riveted ships with paddle wheels?). Other innovations live through decades and even centuries and we know them well – diesel engines, propellers, radio communication. And normally we remember those brilliant pioneers who put their imagination outside the box and created something that no one could think of. Unfortunately, there are some exceptions…

If one looks at the design of modern icebreakers, he/she could notice that a lot of changes have been introduced over the past decades – nuclear power engines, azimuth thrusters, welded steel etc. However, it can also be noticed that the general hull form and traditional icebreaker bow shape remain mainly the same.

Figure 1: The largest to date Nuclear power Icebreaker Arctica (60MW) delivered in 2019 (Photo by Nick Grigoriev)

Figure 2: Another Icebreaker Arctica (55MW). She was the first icebreaker achieved the North Pole (1977).

Figure 3: The first Nuclear powered Icebreaker Lenin (32MW), 1957.

Figure 4: The first Arctic icebreaker Ermack (6MW) 1899.

So, who was the first engineer (or may be not engineer?) who had the bright idea of the innovative bow design capable to go through ice?

2. NEWCASTLE, ENGLAND, 1899

Up until now, a lot of people thought that this man was the famous Russian admiral Stepan Makarov who initiated building the first arctic icebreaker Ermack. This impressive vessel was built in Newcastle, England in 1899 and she had the traditional icebreaker bow shape. Nevertheless, Makarov denied that the idea was developed by him.

Figure 5: Admiral Stepan Makarov (1848-1904)

Admiral Makarov wrote in his book “Ermack in Ice” in 1901: “The idea of icebreakers originated in Russia ... The first person who wanted to fight ice was entrepreneur Michail Britnev ... As you know, Kronstadt navy base is close to St Petersburg and is located on Kotlin island. In summer, communication was organised by boats, in winter by sleighs. But late Autumn and early Spring was the most problematic periods – because of moving ice in Finish gulf the base became literally isolated from the mainland.

So, an extended communication between the St Petersburg and Kronshadt base and delivery the post and goods would bring good money for an entrepreneur. And the man who challenged the natural limitations and spotted the opportunity was Britnev. In 1864, he put one of his boats – Pilot in a dry dock, cut off the bow and attached the new one with a special shape capable to run into the ice and break it off. This little steamer did what seemed impossible: she extended the navigation time in the late fall by several weeks ...” [1].

Figure 6: Michail Britnev (1822-1889)

Makarov also mentioned another interesting story: “In 1871 there was an extremely harsh winter in Europe; the entrance to Hamburg was frozen, and it was decided to build icebreakers. Engineers were sent to Kronstadt to watch the Britnev’s vessel. They bought Pilot’s drawings for 300 roubles, and, in accordance with these drawings, the first icebreaker was built for Hamburg...” [1]. The famous Russian historian Vladimir Andrienko made impressive research about the first icebreakers and published the book “Icebreaking fleet of Russia in 1860-1918’ in 2009 [2]. Vladimir searched for any materials that Makarov could use for his book in 1901. The admiral’s archive is well preserved, but Vladimir could not find any references on where he took his material from. A few articles about icebreakers published in newspapers “The Kronstadt Bulletin” and “The Kotlin” at the end of the 19th century just copy what Makarov said [3, 4].

3. ST PETERSBURG, RUSSIA, 1860s.

Due to the lack of archival materials, the newspaper “Kronstadt Bulletin” remains the main source for reconstructing the history of Britnev’s steamboats. The systematization and generalization of newspaper publications over a 25-year period made by Vladimir, day by day, describing the voyages of private and port vessels in the Kronstadt region, allow us to better understand what was happening at that time.

In the early 60s of the XIX century during the navigation period, communication between Kronstadt, Petersburg and Oranienbaum was carried out by private wheeled steamers. In autumn, at the beginning of ice formation in the Gulf of Finland and on the Neva, the direct connection with the capital was usually ceased. After the formation of the ice cover, pedestrian and toboggan roads appeared on the mainland from the Merchant Harbour. During the period of freezing and ice drift the Kotlin was cut off from the mainland.

The first attempts to maintain communication between the island and Oranienbaum in the ice were made with the help of a port steam vessel Pilot in 1862 and 1863. Judging by the later reports of the newspaper [5], no concrete successes were mentioned, except for rescue operations during navigation.

Mikhail Britnev, the owner of barges, floating cranes and tugboats, decided to rebuild the Pilot’s bow copying the shape of the “ice sleigh”, previously used to make a channel in ice. His hope was that the combination of the special bow shape together with steam power plant and strength of the iron hull would make it possible to ensure safe manoeuvres in ice. The first regular Pilot voyages in ice were reported for the first time in the spring of 1864, and even advertisements of these voyages appeared in the local newspapers at the same time [6]. So, from the spring of 1864 until 1889 (25 years!) when the new icebreaker Luna was delivered, the small Pilot remained an indispensable participant in the communication with Kronstadt in the ice.

Vladimir highlights in his book that it was not an easy service - "Pilot" fell into extreme situations almost every year, repeatedly rubbed in the ice, threw aground, hit ice pieces, but only the rudder, propeller and shaft suffered. The ship was quickly repaired, and she went into operations again.

Of course, risky ice voyages brought considerable profit. A ticket for an ordinary paddle wheeled steamer to Oranienbaum cost 20-30 kopecks. depending on the class (cabin, benches on the deck). Britnev's vessels were used mostly by private passengers (50-60 people on the deck per vessel), and the ticket cost 1–1.5 rubles.

Incidentally, Vladimir says that the Admiral Makarov's story of the creation of the first Hamburg icebreakers was not actually confirmed by the documents. It is known that the Hamburg port Committee announced a tender in 1871 calling for proposals to create an icebreaker. There were 24 projects, and the project of engineer Mr Steinhaus won. The Icebreaker designed by him "Eisbreher I" and all other German river and port icebreakers, had characteristic which were not similar to Pilot [7].

Figure 7: A Soviet post stamp showing possible look of Pilot. It should be noted that the vessel on the stamp has the Russian Navy flag which simply could not be allowed on commercial boats.

Figure 8: Unconfirmed image of Pilot vessel.

Vladimir also mentioned that there are not many details about Pilot, and no one actually knows how she looked like - there was not a single reliable image of the Pilot ship, not to mention the drawings. The "icebreaker" shown on the Soviet postage stamp issued in 1976 is a repetition of the appearance of a nameless steam vessel (or steam barge), depicted in the painting by A.P. Bogolyubov "Opening of the Sea Canal in St. Petersburg".

At the end of the 50s. XX century B. Zylev made a not very successful reconstruction of the "Pilot": a small tugboat with design features that could be attributed to the end of the 19th - beginning of the 20th centuries, but not to the 60–70s. XIX century.

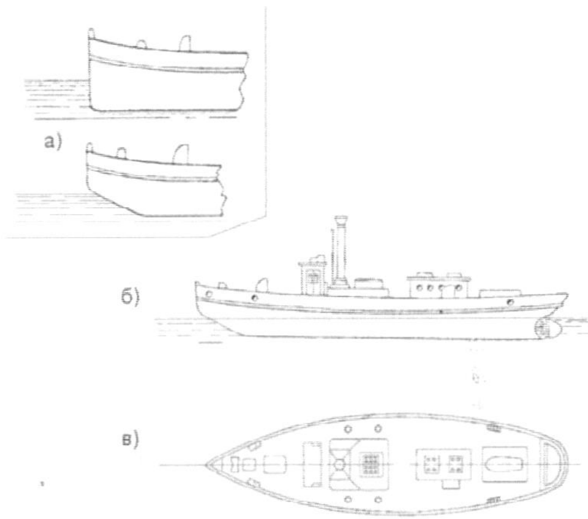

Figure 9: Reconstruction of possible Pilot's look made by B Zyev in 1950s.

Vladimir also found a list of the vessels assigned to Kronstadt port dated the mid-80s XIX century. The Pilot listed in this book has deadweight of just 10 tons which seemed too small for this kind of vessel. Was it a typo or it was another Pilot?

In 1896, Britnev provided the port of Kronstadt authorities with fairly detailed information about his Pilot: power - 25/125 h.p. length - 20.7 m, beam - 3.5 m, draught – 2m. The steamboat's iron hull had 1⁄4 '(6.35 mm) thick metal sheets, and the bow and waterline plating was 3/8' (9.52 mm). The distance between frames was 2 feet (0.61 m) [8]. This vessel was also in the "List of River Steam Vessels of Russia" (for 1897 and 1902) [9] but no further details have been found.

So, Vladimir in his book concludes that despite the regular mention of Pilot in the Russian archives, this ship remains a mystery. So how this historical vessel could look like? Is there a chance we could ever find this out? By the look of it sounds like an impossible task (there is a Russian expression "searching for a needle in the stack of hay"). How many thousands of the vessels of this type have been built over 160 years?

Nevertheless, an interesting detail was found in Vladimir's book. He stresses that it is not clear when and where the vessel was actually built, but he puts a line in a table that that the steam engine was most probably built in Mitchell's factory in Newcastle, England.

4. BACK TO NEWCASTLE, ENGLAND, 1860s

First of all, let us investigate who was Mr Mitchell from Newcastle. Charles Mitchell, who was born in Aberdeen on 20th May 1820, founded the shipyard that bore his name in 1853. He served his apprenticeship with Simpson & Company, iron founders of Aberdeen, before moving to Newcastle in September of 1842. In Newcastle he worked for John HS Coutts, a yard owner, also originally from Aberdeen. Charles worked for Coutts until 1844 before moving to work in London and then travelled extensively in France, Germany and Italy. He then returned to Newcastle in 1852 to set-up his own Low Walker yard next to the Coutts yard.

The yard built over 90 vessels of various types for Russia and Charles Mitchell, together with his business partner Henry Frederick Swan, set up a shipbuilding yard for the Tsarist government at St Petersburg. Several warships were built there under the company's direction. In recognition of his services, Tsar Alexander II made Charles a Cavalier of the Order of St Stanislaus, a rare honour for a British shipbuilder.

Charles Mitchell also had a close working relationship with William Armstrong and Mitchell & Co built many hulls for Armstrong warships. William Armstrong had established a company at Elswick in 1847 and had become one of the world's leading armament manufacturers. This close working relationship developed into a full agreed merger between the two companies in 1882, forming a new company called Armstrong, Mitchell & Co Ltd [10].

Figure 10: Charles Mitchell (1820-1885) [11]

So, it seems Charles Mitchell had a good connection to Russia and according to the records, put together by Shipping & Shipbuilding Research Trust, the first vessels for Russia - BRATETZ "Little brother" (yard number 42) [12] and SISTRITZA "Little sister" (yard number 43) [13] were built in Mitchell's yard in Newcastle in 1857 - just a year after the Crimean war where the British and the Russian actually fought against each other.
If nothing was found in Russian Archives about Pilot vessel, there may be some track records in British archives then?

Apparently, a short line has been found in the Shipping & Shipbuilding Research Trust records [14]

Name: PILOT
Type: Pilot Vessel
Launched: 14/05/1862
Completed: 1862
Builder: C Mitchell & Co
Yard: Low Walker
Yard Number: 103
Dimensions: 23grt
Engines: Steam
Construction: Iron
Reg Number: n/a

It sounded like a promising entry; however, the name Pilot is not a rare name! There may be dozens or even hundreds of small vessels with this name. Would we be sure we are talking about the right Pilot? Apparently, we would! Because there are a few more words in the registry book related to this entry:

History:
1862 MO Britneff, Cronstadt
No later history known

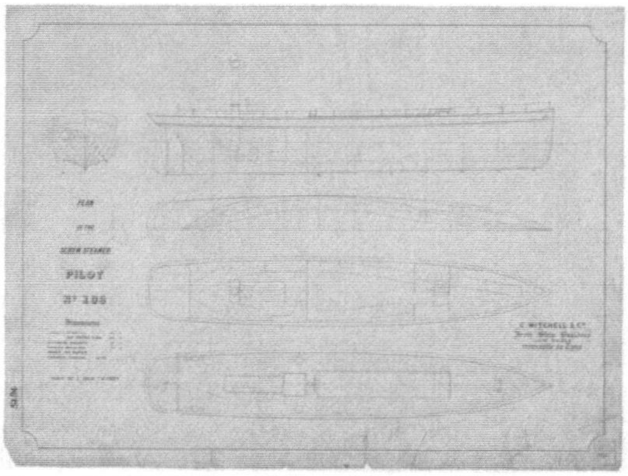

Figure 11: Pilot, General Arrangements, Tyne and Wear Archives [15]

It may be interesting to note that there were two ways of translation Russian names in Europe. The most common that time was the French way (with double "f" at the end) – there are still residuals of that – Stroganoff, not

Stroganov. Nevertheless, in Cyrillic both Britneff and Britnev mean absolutely the same together with Cronstandt and Kronshatdt. So, no doubts that we are talking about the right Britnev and hence the right Pilot then! What does it mean? It means we know the unique yard number of this vessel now – 103.

Then a line of communication was opened with Tyne and Wear Archives in Newcastle upon on Tyne asking whether they have any information about the Yard Number 103 completed at the Low Walker yard in 1862. Everyone would admire the British way of preserving history when such reply is received – The archive has General Arrangements of this vessel in its collection!
Remarkably that after 160 years we can finally see how the first icebreaker looked like.

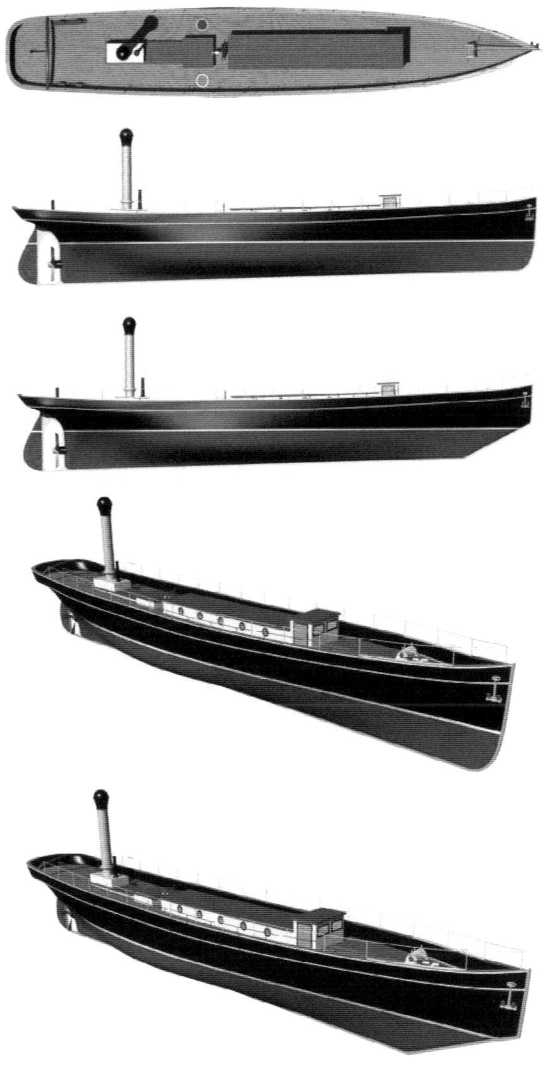

Figure 12: developed 3D model (with and without modified bow) based on the drawings.

5. CONCLUDING REMARKS

It is important to note that sometimes a discovery does not lie in one place - sometimes it is important to bring a few pieces together to match the facts and find the missing link. So, The Tyne and Wear Archives kept the drawings for 160 years, assuming she was just an ordinary small steamer without knowing that she actually was the first icebreaker in the world. The Russian researches on the other hand, knew a lot about the vessel, but for some reasons could not make a connection to British archives. Only bringing these two pieces together helped us to reveal an outstanding story, which is described and presented in here.

And another observation. Despite many people's belief in Russia as a motherland of icebreakers, bearing in mind the famous history of Russian arctic ships, the largest icebreaker fleet to date. It is interesting to note that both the first icebreaker Pilot (1862) and the first arctic icebreaker Ermack (1899). So, despite being Russian ventures they were in fact built in English Shipyards.

5. AUTHORS BIOGRAPHY

Dmitriy Ponkratov is a naval architect with international experience gained in Russia, China, Norway and the UK. He worked as an associate professor at the Norwegian University of Science and Technology and a senior consultant at the world-renowned Technical Investigation Department at Lloyd's Register. Currently, he holds the position of Technical Director at the Royal Institution of Naval Architects.

Jaime Pérez-Martínez holds the current position of Technical Manager at the Royal Institution of Naval Architects. His previous experience includes Project engineer at a Repair Shipyard in Gran Canaria and, in an engineering consultancy company in Madrid. Also, his service in the Spanish Marine Infantry Corps. He is currently pursuing his PhD at the Universidad Politécnica de Madrid.

REFERENCES

1. Makarov S, Ermack in ice, St Petersburg, 1901, reprinted in 2011, EKSMO publishing house.
2. Andrienko V, Icebreaking fleet of Russia 1860 – 1918, Paulsen publishing house, 2009.
3. The first Russian icebreaker, The Kotlin, newspaper, 23 March 1899.
4. The Kronstadt Bulletin, newspaper, 28 May 1889.
5. The first Russian Icebreaker, The Kronstadt Bulletin, newspaper, 26 November 1872.
6. The Kronstadt Bulletin, newspaper, 14 April 1864.
7. Yudashke F, Construction features of icebreakers, Ree Dech Gafen yard, Jan 1929.
8. State Russian Navy Archive, f 427, ch.1, case 173, page 283.

9. The list of steam river vessels of European Russia of 1900, St Petersburg, Issue 4, 1902, pages 212-213.
10. Charles Mitchell, the Shipping & Shipbuilding Research Thrust, http://www.tynebuiltships.co.uk/Mitchell.html
11. Dr Charles Mitchell 1820-1895, Aberdeen's Links with Tyneside Shipbuilding, http://www.mcjazz.f2s.com/TynesideLink.htm
12. TYNE BUILT SHIPS, A history of Tyne shipbuilders and the ships that they built, the Shipping & Shipbuilding Research Thrust http://www.tynebuiltships.co.uk/B-Ships/bratetz1857.html
13. TYNE BUILT SHIPS, A history of Tyne shipbuilders and the ships that they built, the Shipping & Shipbuilding Research Thrust http://www.tynebuiltships.co.uk/S-Ships/sistritza1857.html
14. TYNE BUILT SHIPS, A history of Tyne shipbuilders and the ships that they built, the Shipping & Shipbuilding Research Thrust http://www.tynebuiltships.co.uk/P-Ships/pilot1862.html
15. Yard no. 103, Pilot. General arrangement 1862, Ref No DS.VA/4/PL/2/5296, Vickers Armstrong & Co Ltd Collection 1853 – 1971, Charles Mitchell (later Armstrong Mitchell) ship plans Section 1853 – 1907, Tyne and Wear Archives.

GALATEA II: REBORN OF A CLASSIC

M. Castro Ruiz, R. Perez Fernandez, Universidad Politécnica de Madrid, Spain

SUMMARY

The history of the **Galatea** dates back to the 19th century when in 1896 she was launched and Christianised as **Glenlee** at Glasgow. She sailed as a merchant ship in the United Kingdom and Italy during the first period of her operational life, and was later adapted with mechanical propulsion.
After several circumnavigations, in 1922 she was renamed **Galatea** to serve in the *Spanish Navy*, where she remained in service for 60 years. Since 1993 she has been resting in her hometown in Scotland as a museum ship.
As a tribute to the extensive and remarkable history of this ship, and in order to recover and preserve the naval tradition in Spain, it has been proposed to design a ship with the same morphology as the **Glenlee** to stoke the spirit of the **Galatea** and inspire the construction of new sailing ships. She will be adapted as a military training ship of the *Spanish Navy*. Therefore, studies in stability, propulsion, general arrangement and structural calculations will be necessary to validate the transformation of the **Glenlee** into a new **Galatea II** complying with mandatory regulations and technological advances that will encourage its operation to the future naval officers of the *Spanish Navy*.

1. INTRODUCTION

Over the last century, sailing has become part of nautical activities and recreation. However, its use as a type of propulsion has endured for many centuries, which suggests that mechanical propulsion is actually pretty new.

However, it is still possible to find many classic sailboats of varied size sailing through seas around the world. The art of classic sailing requires a high physical and mental strength, so instruction on "sail training" is a highly valued skill in the merchant marine as well as in the military academies.

*Figure 1: **Glenlee** at the Riverside Museum on Pointhouse Quay. Photo taken by R. Perez in June 2011*

During this paper it will be developed a brief academic research as regards the **Glenlee** (see *figure 1*), she is a steel-hulled three-masted barque, built as a cargo ship at Port Glasgow under that name in 1896. After several shipowners´s changes, she became the sail training ship **Galatea** in the *Spanish Navy* in 1922. Since 1993, carrying her original name, **Glenlee** [1] has been a museum ship at the *Riverside Museum* on Pointhouse Quay, Glasgow, known as *The Tall Ship* at Glasgow Harbour.

During the following paragraphs, there is going to be described a state of the art of tall ships.

1.1. CIVIL AND NAVAL SAILING SHIPS

Outside the military sphere, training ships are part of the nautical sector of training and recreation. This activity considered professional is called "sail training" sailing and, although it seems to be unknown in Spain, it is very popular from northern Europe countries like Norway, Holland and Denmark.

In Spain it is difficult to find this type of activity related to these classic sailboats, but it is true that there are several associations and foundations that fill their decks in summer with volunteers from all over the world to participate in regattas internationally and at maritime festivals throughout Europe.

Civilian and naval training sailing ships have great similarities in terms of hull forms. However, when operating in different civil and military fields, the configuration of their spaces is completely different, since the military regime is in all respects more demanding both in capacities, organization and discipline. For example, in the case of a training ship commissioned for military training is important to serve the compartmentation of spaces depending on the officer's scale, while on a civilian training ship this division is not really necessary.

However, all these ships participate annually in different events and regattas bringing together a large number of crews from around the world, such as the *Tall Ship Race*, one of the best-known international regattas. In this case, ships of different navies such as the **Esmeralda** or **Juan Sebastián de Elcano** compete with civil sailing ships such as the **Gulden Leeuw** or the **Pelican of London**. That is, there is a bond of union among all training ships, and is to maintain and expose that authenticity that characterizes them.

1.2. SECOND CHANCE TO SAILING SHIPS

Not so long ago, merchant sailing ships competed to transport their cargo around the globe as fast as possible. Time has always been a key factor in the success of a transport operation. So much so that regattas were held annually to find out which sailboats were the fastest of them all.

With the opening of large canals such as the *Panama Canal* or the *Suez Canal*, and the industrialization period of the early nineteenth century, sailboats were gradually replaced by steamships, with mechanical propulsion.

Therefore, the construction of this type of ships was drastically reduced and many were scrapped or sold for parts, depending on their construction material: wood or steel. Very few have survived over the years, as their high cost of maintenance hampers their maritime activity.

However, some of them were transferred to different ports as ships museum in memory of such characteristic and unique vessels. Others were rescued as training ships for military academies in different countries or as pontoons. Sailing is a demanding discipline, highly valued for the correct training of cadets and midshipmen from around the world.

Although there are navies with almost centenary sailing ships, such as the **Juan Sebastián Elcano** in Spain or the **Amerigo Vespucci** in Italy, many countries chose to build new training ships, but with the same look as olders, like the **BAP Unión for Peru** or the **Bima Suci** for Indonesia.

1.3. SAILING SHIPS IN SPAIN

Although the preparation of the students of the different navies has evolved according to their needs and technological advances, training sailing ships are part of their training for almost two centuries.

In the case of the *Spanish Navy*, the appearance of the first training ships dates back to 1871 with the inauguration of the "**Asturias**" floating school. Originally, this ship was a frigate called "**Princesa de Asturias**", launched in 1857 and named frigate of propeller to have mixed propulsion: sail and engine. At the end of its operational life, she's anchored in La Graña, Ferrol, where she would be converted to a floating school. The two frigates **Blanca** and **Almansa**, also called floating schools, would succeed her.

It was not until 1886 when the *Spanish Navy* bought a clipper for 60,000 pesetas (roughly 360 EUR) built in Glasgow, called **Carrick Castle**. She would be renamed **Nautilus** and would be the first real training ship, since previous ships did not sail: hey simply remained moored at the military naval academies. That was also when the *Spanish Navy* used training ships not only for academic purposes but as floating embassies as well. Proof of this is the **Nautilus**, being the first Spanish ship to visit Cuba after its independence.

Nautilus would be replaced by the Galatea, also an English ship, and later by the **Juan Sebastián de Elcano**, see in *figure 2*, current and only training ship of the *Spanish Navy* built in the Cadiz shipyards of *Echevarrieta y Larrinaga*.

*Figure 2: Tall Ship **Juan Sebastián Elcano**. Source: Spanish Navy*

As of today, the **Juan Sebastián Elcano** is the third largest training ship in the world along with its twin, the Chilean training ship **Esmeralda**. It is also one of the most long-lived, as it will be 100 years old in 2025, after traveling more than 1,700,000 nautical miles and more than ten times around the world.

2. FROM GLENLEE TO GALATEA

The predecessor of the current *Spanish Navy* training ship is the **Galatea**, as it can see in *figure 3*, originally called **Glenlee.**

*Figure 3: **Glenlee** as Spanish **Galatea** in an old picture*

The **Glenlee** was built in Glasgow in 1896 at the Bay Yard Shipyards, Scotland. It was one of ten sailboats from the same design, commissioned by Archibald Sterling & Co. Ltd. She was designed as a merchant ship for the transport of grain between the United Kingdom and the Pacific islands such as Australia and New Zealand. Two years after its acquisition, she was bought by another shipping company called *Islamount Sailing Ship Company Ltd.* She was then renamed with the same company name, **Islamount**, and continued sailing as a merchant ship on the same routes, to the Pacific.

Later she would be bought by another shipping company called *Richard Thomas & Co.* from Liverpool, in 1905, until she was put in the service for the *Royal Navy* during the First World War. Her first destination then was Buenos Aires, Argentina, to take supplies. Then she crossed Cape Horn, one of the most demanding waters to navigate. During the war she remained active in South American waters in different countries, until 1918 [2].

A year later she was acquired by the Italian shipping company Stella, where she was renamed **Clarastella**. Until 1922 she operated again as a merchant ship in the Mediterranean Sea, until the *Spanish Navy* bought her.

Through a new law in Spain to acquire sailboats, the **Clarastella** changed her name again: **Galatea**, in honour of the famous nymph from Greek mythology. By then, this sailboat had travelled many nautical miles accross the seven seas.

Figure 4: Original General Arrangement of Spanish Navy ***Galatea***

Upon arriving to Spain, she underwent an internal transformation to be adapted as a training ship for the *Spanish Navy* cadets, with a proper naval configuration, as it can be seen on *figure 4*. Its rig, originally designed as a barge, did not suffer any considerable changes. However, the main deck was modified to install a new command bridge and the poop was extended to install the lifeboats. Fixed ballast was also added to lower the position of the centre of gravity and thus improve the stability and seaworthiness of the sailboat. In addition, it was equipped with four six-inch guns.

For a short time, it sailed alongside its predecessor, the **Nautilus** training ship, and its successor, the **Juan Sebastián de Elcano**. However, the latter did not take long to relieve it of its functions as an officer´s training ship, and after a series of changes in its engine room, the **Galatea** was assigned as a manoeuvre training ship.

In 1936, the Galatea suffered another armed conflict: the *Spanish Civil War*, and even later there would be a third one: The *Second World War*. Although she had some unfortunate encounter with a German submarine, there was no incident to regret. As a curiosity, the **Galatea** was then equipped with several Spanish flags on the tacks and fins to be recognized as from a neutral country in the war.

As a training ship, she also made transoceanic voyages, visiting ports on the American continent on several times. During her life as a training ship she suffered some severe storms that on one occasion left her without sails, with a heel of 42° and the engine-room flooded.

Her last trip was at the end of 1959. Later she was moored in Ferrol as a manoeuvring pontoon until she was decommissioned in 1982.

Since then, her condition has gone from bad to worse. Its deterioration was remarkable, and the date of its scrapping was only delayed due to the complaints of the fervent followers of the **Galatea**.

Several Spanish cities were interested in the ship and its recovery. Finally, it went to Seville to be transformed into a communications centre for *EXPO 92*, next to the *Torre del Oro*, to later be converted into a naval museum.

Unfortunately, after her transfer to Seville, the high cost of her repair caused it to be abandoned in the basin of the Gualdaquivir river, being plundered by scrap dealers and collectors and frequented by beggars as a refuge. After several fires and waterways, she ended up perched on the river bed.

Figure 5: Mezzanine deck of the ***Glenlee*** *at the aft end. Photo taken by R. Perez in June 2011*

She was refloated to prepare for scrapping. She entered a public auction, which was attended by the Scottish association *Clyde Maritime Trust*, who acquired the ship in 1992. She was towed a year later to Greenock, and then to Glasglow (as it can be shown in *figure 5*), where it was received by the Mayor of the city and the Consul of Spain.

With a budget of £2.1 million she was fully restored and converted into a museum ship on the banks of the Clyde River [3].

As regards her generating plant, during the Galatea years, she was fitted with two 100 kW dynamos each driven by two 160 HP Ruston Hornsby 3 cylinder diesel engines. They generated electricity at 110 V DC which was controlled by the switchboard.

*Figure 6: **Glenlee** marble switchboard panel. Photo taken by R. Perez in February 2018*

The marble switchboard panels, see *figure 6*, were cleaned of dirt deposits and grime that had accumulated over the years which Galatea had spent on the bottom of the Guadalquivir River. The cleaning revealed the few remaining circuit labels which, together with Spanish archive notes, help us to understand how electrical power was used on board.

3. OLD MERCHANT SHIP AS TALL SHIP

The **Galatea** arrived to Spain in 1922 after passing through three different shipping companies. For the first time in her history, this merchant ship performs her function as a training ship, so her space is reconditioned for this purpose.

A few years later, in 1925, the **Juan Sebastián Elcano** was built in Cádiz, being then one of the largest sailboats in the world with 4 masts and 113 m in length overall.

The **Galatea** was in the background as a training ship for sub-officers and midshipmen of the *Spanish Navy*.

Given the short time she operated as a training ship, readers could wonder: was the **Glenlee** a good buy as a training ship? Even being a ship admired by its midshipmen, was she a good sailboat for her new mission?

3.1 DATABASE DEFINITION

Before answering this ambiguous question, it is necessary to know the rest of the **Galatea** competitors, that is, what the training ships of other Navies are like in order to define the characteristics in common among all the sailboats of this class.

This database to develop includes ships from the late nineteenth century to the present day, the latest addition to the list of school ships being the **BAP Union** of the *Peruvian Navy*.

The data obtained is summarized in the following *figure 7*, according to the variation of length overall (m):

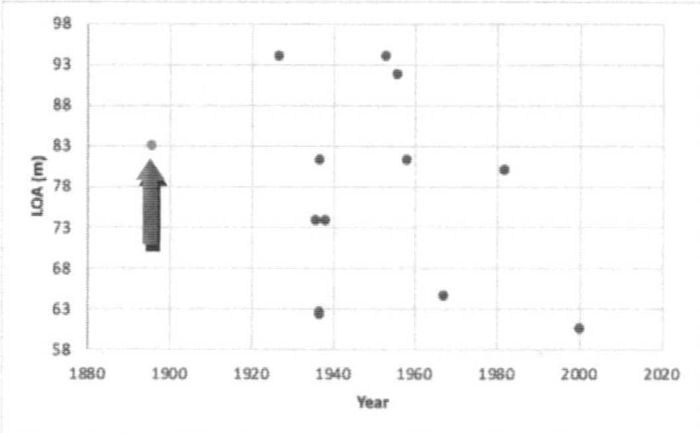

Figure 7: Comparison between length and commissioned year

From this graph it can be seen that the **Galatea** has a length overall above the average, as it happens with ships such as the **Juan Sebastián Elcano** and the **Esmeralda,** two of the largest training ships in the world.

Let's see if the same happens with the displacement in *figure 8*:

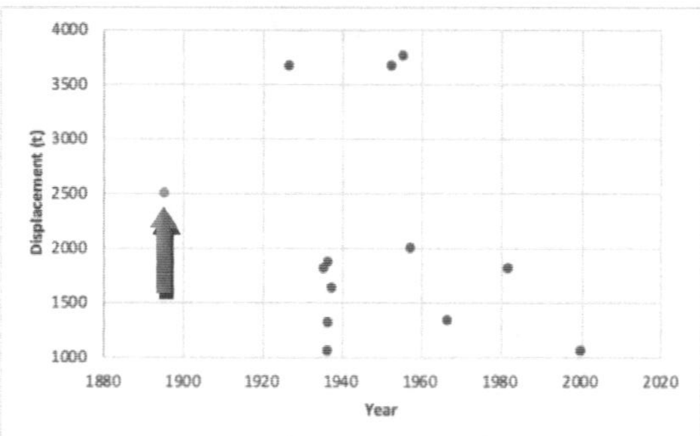

Figure 8: Comparison between displacement and commissioned year

As with the length, the **Galatea** has a greater than average displacement. To give a clear and simple example, the **Galatea**, depending on its draft, could displace up to 2,700 t, while this displacement value corresponds to a larger ship, such as the **Dar Młodzieży**, with 110 (m) in length compared to 83.07 m on the **Galatea**.

Considering this last-mentioned characteristic, the block coefficient of the vessels in the database is calculated. The mean is around 0.48, considering some of the largest ships. The **Galatea** exceeds this average with a block coefficient near to 0.53.

Most of the ships in the database, emphasizing the more modern ones, were designed as school ships with slender shapes to improve their seaworthiness. That is, emphasizing its need to create fast sailboats, with spaces optimized for life aboard the military crew and also fulfilling its representative function of the flag country as embassies.

In other words, none of these training ships was previously a merchant ship. The **Galatea** is a cargo ship converted into a training ship, but its forms were not designed for that purpose.

This results in more round shapes, with a slightly cylindrical body and with a lot of storage space that would later be used for the decks where to accommodate the midshipmen.

Her speed is also affected by this factor, and comparing the speeds of the ships included in the database, most reach an average of more than 12 knots, while the **Galatea** could hardly manage to make more than 10 knots.

3.2 ARE THERE FAST MERCHANT TALL SHIPS?

Like the **Glenlee**, there were numerous cargo sailboats with similar characteristics that have survived till our days. For example, the oil tanker **Falls of Clyde** sailboat, one of the sailing ships with the same design as **Glenlee**. They also share a new function as museum ships. **The Falls of Clyde** remains moored in Honolulu, Hawaii [4].

Hull shapes for sailing ships evolved from being relatively short and blunt to being longer and finer at the bow. By the 19th century, ships were built with reference to a half model, made from wooden layers that were pinned together. Each layer could be scaled to the actual size of the vessel in order to lay out its hull structure, starting with the keel and leading to the ship's ribs. The ribs were pieced together from curved elements, called futtocks and tied in place until the installation of the planking. Typically, planking was caulked with a tar-impregnated yarn made from manila or hemp to make the planking watertight. Starting in the mid-19th century, iron was used first for the hull structure and later for its watertight sheathing.

It would be wrong thinking that all sailing merchant ships have a large beam, rounded shapes, a cylindrical body and perform slower than other ships.

Figure 9: Cutty Sark, now a museum ship and tourist attraction in Greenwich, Southeast London. Photo taken by R. Perez in January 2017

One of the best examples that can be exhibited either for her almost legendary fame or for her good seaworthiness is the **Cutty Sark**, built in 1869 as a tea-clipper, at the River Leven, Scotland in 1869 for the Jock Willis Shipping Line. She was one of the last tea clippers to be built and one of the fastest, coming at the end of a long period of design development, that halted when sailing ships gave way to the steam propulsion (see *figure 9*). This sailboat was also a tea and wool goods ship in its passage through several shipping companies. In addition, she was the fastest cargo sailboat of her time.

Figure 10: City of Adelaide, now in the middle of a restoration process in order to become a museum ship, Port Adelaide, South Australia. Photo taken by R. Perez in October 2016

City of Adelaide, see *figure 10*, is the world´s oldest surviving clipper ship, among the only two that survive, being the other one the already mentioned **Cutty Sark**. With **Cutty Sark** and **HMS Gannet** (built 1878; a sloop-of-war in Chatham as it can be seen in *figure 11*), **City of Adelaide** is one of only three surviving ocean-going ships of composite construction to survive.

The clipper type sailboats (current **Stad Amsterdam, Cisne Branco, Padua and Shabab Oman II**) appeared in the 1930s in Scotland and are characterized by their elongated and slender shapes and their high speed, which allowed them to compete with steamboats. To this day, new clippers are still being projected, such as the *Brazilian*

Navy training ships, the **Cisne Branco**, or the **Stad Amsterdam**.

*Figure 11: **HMS Gannet**, a training ship in 1903, restored in 1987, and now part of the UK´s National Historic Fleet. Photo taken by R. Perez in October 2017*

In other words, the amount of cargo to be transported and ship's speed was optimized over the years and under the needs of each owner. As today, it is possible to find faster cargo ships with less cargo and vice versa.

Nor is the **Glenlee** the only merchant ship that operated and operates today as a military training ship. A clear example of a large merchant sailing ship converted into a training ship is the **Padua** (renamed **Kruzenshtern**). It is the third largest training ship in the world, and also one of the fastest. The reason? It is a clipper type sailboat that reaches more than 17 knots of speed.

4. HULL ANALYSIS

Now that the characteristics of the **Galatea** against other active training ships and sailboats are known, it is the time to proceed for analyzing her forms.

Table 1: Hull forms main dimensions and speed

Length overall	94.57 m
Hull length	83.07 m
Length between perpendiculars	74.87m
Width	11.41 m
Depth in hold	7.45 m
Draft, fore	5.2 m
Draft, aft	6.2 m
Sail area	2,800 m^2
Displacement	2,500-2,900 t
Speed (propelled)	8 kn

For this, it is necessary to start from its original dimensions shown in *table 1*, that is, those corresponding to the **Glenlee**. The only significant difference from the **Galatea** is that the original had smaller sail area.

Through the Rhinoceros® 3D design program, the shapes of the **Galatea** hull are modeled using as a reference the drawings provided by the *Logistics Support Headquarters* of the *Spanish Navy* in Madrid.

The result of these forms is the following shown in *figure 12*:

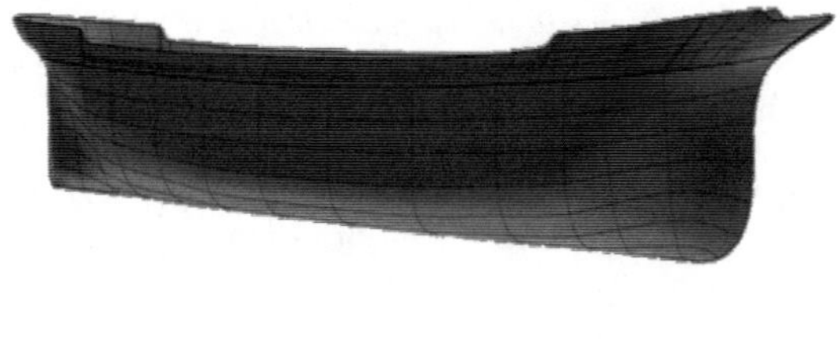

*Figure 12: 3D model of **Glenlee/Galatea***

On both merchant sailboats and sailboats with finer lines, the bow and stern shapes usually look similar.

The shape of the bow is thin and slender and its sections are V-shaped, reducing wet surface. Given the risk that these types of vessels suffer from being kicked, it is very likely that the effect of slamming will occur, so these forms favor the dissipation of the impact of the waves against the bow.

The stern, in this specific case, has similar style to the rest of the sailing ships that it is possible to find today, called the cruise stern type, and it shown in *figure 13*.

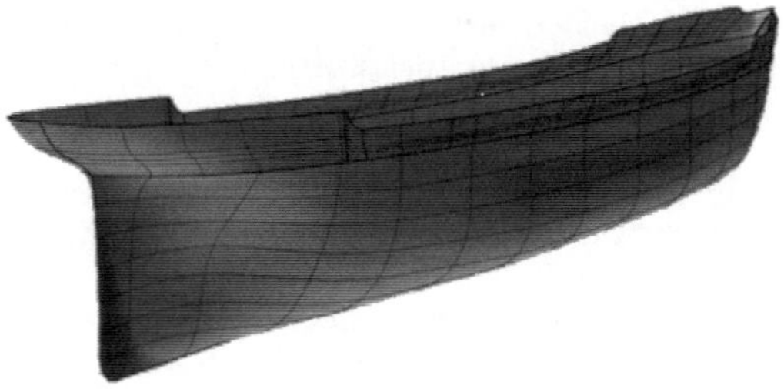

Figure 13: 3D detail of the stern

By testing the hull forms with the software Maxsurf®, it will be obtained the information regarding their hydrostatic values and the areas curve for the design draft estimated at 5.6 (m).

According to the areas curve (see *figure 14*), both fore and aft, fine shapes favor the water flow in these zones.

It is also possible to observe the slightly cylindrical body that this curve (helmet-type) presents, resulting from a design to store large amounts of cargo. Other type of sailboats, especially smaller in length, would show a curve of areas without a plateau in the central body.

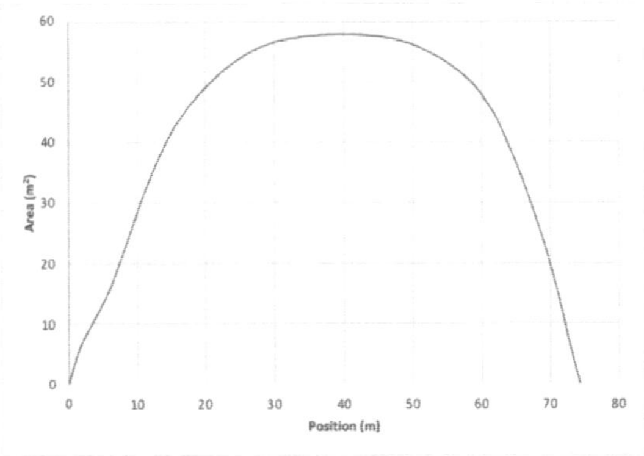

Figure 14: Area Curve graph

4.1 HYDROSTATIC DATA

For the estimated design draft of 5.6 m, the displacement corresponds to approximately 2,800 t.

The hull forms coefficients obtained from the Maxsurf Modeler simulation are as follows in *table 2*:

Table 2: Hull forms main coefficients

Block coefficient	0.61
Prismatic coefficient	0.73
Maximum section area coefficient	0.84
Waterplane Area coefficient	0.79

The resulting value for the block coefficient is high compared to the mean obtained from the database, of approximately 0.48.

This value confirms that the shapes of these merchant ships were, indeed, less slender than the more modern fast ships.

Consequently, the coefficients prismatic, maximum section area and waterplane area are therefore high relative to the mean.

5. CONCLUSIONS

The classic sailing ship **Glenlee** travelled the world as a merchant ship and military training ship. She is a veteran that today is a large floating museum.

As it was stated during the paper, during the transition from UK to Spain, except for the hull a new vessel had to be reconstructed. All the changes made by the *Spanish Navy* had to be removed, such as all the cabins built for the trainees and a lot of scrap iron ballast in the frames of the holds. First of all she was given back her original name, **Glenlee**, by the *Lord Provost of Glasgow* on 6 of July 1993 when the ship arrived in Glasgow for the first time since her launch in 1896. **Glenlee** is now recognized as part of the *National Historic Fleet*.

Besides being a museum ship and tourist attraction, **Glenlee** should be revisited from a technical and historical points of view.

6. REFERENCES

1. NATIONAL HISTORIC SHIPS, *"Glenlee. Delivering the goods"*, Website, Unknown
2. BROWN, P. *"Britain's Historic Ships: A Complete Guide to the Ships that Shaped the Nation"*, Book, 2009
3. SCHAUFFELEN, O. *"Chapman Great Sailing Ships of the World"*, Book, 2005
4. NATIONAL HISTORIC SHIPS UK, *"Glenlee"*, Website, Unknown

7. AUTHORS BIOGRAPHY

María Castro is a final year MSc student in *Naval and Ocean Engineering*. Her thesis is based on the design of a classic sailboat that complies with current regulations and meets the needs of a training ship for the *Spanish Navy*. She currently works at *Navantia* in *International Business Management Department*.

Rodrigo Perez, Ph.D. is the head of *Naval Shipbuilding Business Development* in *SENER Ingeniería y Sistemas S.A.* He is also professor in several subjects at the *Marine Engineering School* of the *Technical University of Madrid*.

THE *ROYAL LOUIS* (1668), A SAILING MASTERPIECE

C. Tacchella, University of Genoa, Italy

SUMMARY

The French Navy's finest hour began in the seventeenth century, when Colbert became the Prime Minister and decided to improve the national shipyards. Despite poor starting conditions, the results of Colbert's efforts led France to be one of the major players in the struggle for naval power. In that moment, the European shipbuilding had a changing in construction techniques mostly in warships. In France, this changing also involved vessels aesthetics. Indeed, shipwrights' tasks were not only to build ships with great naval skills, but vessels had also to be aesthetically striking and eye-catching in order to show at the world the Roy Soleil's power and wealth. Many vessels were built in that style and the *Royal Louis* was one of the largest vessels built in 1668 at the Toulon. With particular attention to the *Royal Louis*, the essay aims to analyze several aspects of the world surrounding those floating masterpieces.

1. INTRODUCTION

European shipbuilding underwent a transformation in the seventeenth century as a consequence of new ocean trade becoming increasingly important. Stronger and larger vessels were required to sail the oceans. Moreover, the arising value of commercial trade towards other continents had led to a new perception of fleets' usefulness in European minds [1]. Indeed, with the new establishment of ocean-based trade, fleets were no longer used only to defend countries' borders and support military campaigns; Navies had become the main means to protect States' commercial routes and achieve economic supremacy. Having a powerful Navy was essential for a State looking to excel in the seventeenth century world. By the eighteenth century this new conception of the purpose of a fleet had become fully established. As testified by Jean-Fréderic Phélypeaux, first Count of Maurepas (1701-1781) and Secretary of State for the Navy, who, in 1745, wrote: «*Le commerce fait la plus grande richesse et conséquemment la puissance des États ... les forces maritime sont absolument nécessaire pour les soutiens du commerce*» [2]. In the 17th century, thus, naval warfare was no longer subordinated to land operations, and thus became the focus for captains, officers and admirals aiming to outline an efficient combat tactics that exploited all the possibilities offered by the artillery. Indeed, these weapons began to be actually useful in naval wars only at the beginning of the sixteenth century. This is when the introduction of gunports allowed ships to carry a greater number of guns. As testified by the well-known *Henry Grace à Dieu* (1514), or *Great Harry*, which boarded more than 180 pieces of artillery (43 cannons and 141 light guns) [3]. The broadside armament highlighted the inferiority of short-range weapons and led to a volume reduction of quarterdeck and forecastle, traditionally used by archers and harquebusiers. The English Navy undertook this paradigmatic change in naval combat, recognizing the future of sea warfare was in heavy cannons. To the point that, in 1618, the *Commission of Reform* recommended that Navy rearrange the artillery on the supremacy of cannons: «*Experience teacheth how sea-fights in these days come seldom to boarding, or to great execution of bows, arrows, small shot and the sword, but are chiefly performed by the great artillery breaking down masts, yards, tearing, ranking and bilging the ships, wherein the great advantage of His Majesty's navy must carefully be maintained by appointing such a proportion of ordnance to each ship as the vessel will bear*» [4].

A different way of fighting was the outcome of this new perspective and the so called *battle line* was written down for the first time on 29 March 1653 by the *Generals at Sea* Robert Blake (1598-1657), George Monk (1608-1670) and Richard Dean (1610-1653), who drafted several norms contained in the *Sailing and Fighting Instructions* [5]. The biggest problem in naval battle planning was the ships heterogeneity. Thus, what fleets needed was a standardizations in shipbuilding in order to increase the efficiency of naval operations and simplify of the organisation depending on the dimensions and the number of cannons their carries. For this purpose, a specific classification was created which divided ships into different rates and all vessels built under that classification were called *ships of the line* [6]. *First rate* vessels were the biggest and most heavily armed ones with a fire power of 70 to 100 cannons or more; these were usually flagships and leaders in naval conflicts. *Second rate* vessels had fewer cannons (generally 50-70) and even fewer on vessels of lower rates.

This changes in shipbuilding influenced almost all of the European fleets and perhaps saw the greatest results in England and the Netherlands. Nevertheless, France developed a peculiar aspect more than other States: their concern for aesthetic ostentation.

2. WHY DID AESTHETIC CONCERNS PREVAIL IN FRANCE ?

It is questionable why aesthetic had a stronger influence in the French Navy than in the others.

Considering that Navies had become state-owned tools, the European political situation has to be taken into account. The consolidation of States' power and the creation of state Navies led to conflicts that were increasingly linked with a growing citizens' sensibility of identity. It is in this sense that the desire for supremacy was no longer just meant as a territorial conquest but also as an ideological phenomenon. The Sovereign's power, and thus metonymically that of the State, was becoming more and more symbolic and significant aspect that had to be transmitted to the rest of the world in every possible way.

Figure 1: Detail. Portrait of the *Royal Sovereign*. 1661 (?). By Willem van de Velde. Source: National Maritime Museum, Greenwich, London, Caird Collection.

The seduction of aesthetics influenced all the main European Navies. The Swedish *Vasa* (1627) and the English *Sovereign of the Seas* (1637) [7] are two of the main examples. However, these magnificent vessels mainly had to do with size and gun-power rather than aesthetics. When in 1634 Charles I Stuart (1600-1649) decided to proceed with the construction of the *Sovereign of the Seas* (or *Royal Sovereign*) [Figure 1], Europe was experiencing a particular time characterized by the increasing of tonnage as a common trend. Almost all the Navies were working on the construction of *three-decker* vessels. The reason for this trend can be explained quite easily, since bigger and sturdier vessels were able to carry more cannons, and thus increase the chances of winning in *battle line* conflict. Even the Dutch Navy, which traditionally had been privileging small and fast ships, was building bigger ships in order to compensate for their small fleet and also to be highly competitive on ocean trade [8]. France started to built the *Couronne* in 1629. It was one of the biggest vessels of its time, 165 feet long (50.3 metres) and boarding 20 heavy guns and others light ones [9]. In this context, Charles I was almost forced by the tough competition to order the construction of a large vessel. The task was entrusted to the *master of the shipwrights* Phineas Pett (1570-1647), who had previously built the *Prince Royal* (1610). That project encountered some opposition. The *Corporation of Trinity House of Deptford* argued that such a vessel would have been unusable and also dangerous due to its size. The *Sovereign of the Seas* was launched in 1637 despite the protests; she was 232 feet long overall (70.7 metres), her structures were lavishly decorated and she became the King's pride. The vessel also performed well in some military campaigns during the Anglo-Dutch Wars. And she was particularly appreciated for her success during the battle of La Hougue (1692). But bulky superstructures immediately showed their problems, increasing the weight and causing issues with the manoeuvrability and stability of vessels. Therefore, in 1651 the *Sovereign* was put in Chatham's dockyard and her superstructures were cut down [10]. Thus, England soon followed the Dutch example decreasing the volume of decorations.

2.1 THE POWER OF DECORATIONS

Among the European States, France was where the artistic character took a central and crucial role in shipbuilding process. Here, decorations excelled and exceeded with ornaments, wooden statues and gold, as never seen before.

Since greatest vessels were used by monarchs as propaganda tools, the main thrust to use ornaments in abundance was bound to come from them. The seventeenth century coincided with a particular king in the France's history, that is Louis XIV (1638-1715). He ascended to the throne in 1643 and during his reign he strove to concentrate power in his own hands, strengthening his political identity trough a programme that is best exemplified by his well-known motto *«l'État, c'est moi»*. The implications of this plan were not marginal. In this context, France was no longer only a territory to be governed and the King did not have to act *on behalf of* the State because he *was* the State. By means of this ideological overlap of State and King, every display of France's power was directly a glorification of the King. Similarly, promoting the monarchy's magnificence and wealth meant glorifying the State. This propagandistic action was extended to all fields.

The Navy was strongly influenced by this approach because vessels became ambassadors of the King and France at sea. Their names were affected too, as testified by changes in the types of names that were chosen. Indeed, until the early years of the 17th century, ships were usually named after saints; the *Saint Michel* (1621), the *Saint Jean* (1621), the *Saint François* (1625), the *Saint Charles* (1628), the *Saint Thomas d'Aquin* (1642), the *Saint Jacques de Dunkerque* (1643) are all good examples of this trend. But during the Sun King's reign, vessels' names were chosen with great attention in order to express royal identity. Depending on the vessels' class, there was a difference in meaning; at the top, there were the two flagships which, given their great importance, were named after the King himself, *Soleil Royal* (flagship in the Eastern fleet) and *Royal Louis* (flagship in the Western fleet). Further down in the pecking order, there were the *First Rate* vessels, typically boarding up to 100 cannons; their names had to refer to royal icons and symbols such as the *Royal Dauphin* (1671), the *Monarque* (1669), the *Couronne* (1669), the *Sceptre* (1671) and the *Lys* (1691). The vessels which were classified in the *Premier Rang Ordinaire* (boarding about 80 guns) and those in the Second Rate (60-70 guns) generally had names symbolizing the Crown's virtues, as the *Superbe* (1690), the *Glorieux* (1678), the *Magnifique* (1685) and the *Invincible* (1691) [11].

Knowing the great importance given to names, one can easily understand that the external design could not be overlooked. Thus, the shipwrights' task was not only to build efficient vessels, but also to make sure they looked astonishing. Following this trend, aesthetics became increasingly important in shipbuilding; to the point that the overabundance of carving led in some case to a decrease in seaworthiness.

2.1 (a) Pierre Puget and *La Reine* (1647)

The concern for aesthetic appearance existed before monarchy started using vessels as means of propaganda. In the first half of seventeenth century, Colbert had not yet embarked on his quest to reorganize and improve the Navy. By that time, the French shipbuilding was held by rich merchants and, in case of need, the State could rent their vessels. Those private ships did not luck of decorations. Ornaments were used for both decorate vessels and show off owner's wealth. The Second Rate *la Reine* was one of these vessels. She was built at the arsenal of Toulon from 1645 to 1647. And the *Grand Admiral* of the French Navy Jean Armand de Maillé, Duke of Fronsac, Marquis of Brézé (1619-1646) called the young artist Pierre Puget (1623-1694) to design the workings drawings of decoration [12]. Calling a specific artist was already in itself a sign of the importance of aesthetics in shipbuilding. But the *Reine*'s history is also interesting for the great ferment she caused.

After her launch, public opinion split in two opposite sides. On one hand, there were the ones who praised Puget for having taken the first steps towards the evolution of the French ships decoration and having introduced luxurious projecting galleries in the stern. On the other hand, there were people complaining about the excessive opulence of the carving, at the expense of seaworthiness.

The stern was the most decorated part. Here, there was a gallery surmounted by a big medallion, which was in the centre and was held by two caryatids. This medallion had inside a representation of Anne of Austria (1601-1666), who was the *Grand Maitrise de la Navigation* [13] from 1646 to 1650. In actual fact, however, the decoration of this vessel was not particularly excessive if compared to the style of that time and it certainly was not on the level of the great magnificence of ornaments which the French shipbuilding achieved over a period of twenty years. Even wooden statues were not totally unprecedented in shipbuilding, as shown by the presence of the sculptor Nicolas Levray (Unk-1678) [14] at the Toulon's arsenal since 1639. And it was he to create the statues of Puget's drawings for *La Reine* [15]. Nevertheless, criticism surrounding this vessel is really interesting because it shows two different schools of thought in France and tells us that there was an early debate about the excessive use of decoration in warships. If, on the one hand, ornaments were appreciated for increasing the beauty of vessels, on the other hand, the undeniable reduction of seaworthiness could not be overlooked for much longer.

For the next twenty years, increasingly large ships were built. On them, artists had the opportunity to display their ability. The *Royal-Louis* and the *Dauphin-Royal*, which were built simultaneously in the 1667 at the Toulon's arsenal, are two of the best examples of this trend.

In the late 1660s, the concern for aesthetics achieved the highest level and that led to a stronger opposite reaction. According to this opinion, the extensive use of carving was useless and even dangerous, since it overloaded both fore and aft parts causing instability. Moreover, the bulky volume of the aft superstructure generated a resistance to the motion of the hull, reducing the vessel speed. And, as important, the realization of ornaments and decorations often caused delay in the completion of ships. Furthermore, all those wooden statues increased the risk of fire started by enemy fireships. For these reasons, many captains of lower rate vessels, which were not under the watchful eye of the monarchy, often jettisoned many ornaments, preferring to sail a good ship than a beautiful one [16]. Nevertheless, despite complaints from captains and admirals, Colbert was sure of the perfection of French shipbuilding and its decoration.

2.2 COMPARISON WITH FOREIGN SHIPBUILDING

Since the 1670s, the belief of Colbert was shaken. The many vessels built with French traditional shipbuilding had started to show their flaws. And so it was that the Minister sent his councillors to foreign shipyards to understand their secrets and compared the different techniques. The men choose for this task were Étienne Hubac (1648-1726), Pierre Arnoul (1651-1719) and Jean-Baptiste Antoine Colbert, Marquis de Seignelay (1651-1690) [17]. All reports demonstrated that differences from French, English and Dutch shipbuilding were not only about decoration. Indeed, French hulls shapes were not the same of foreign ones. The Dutch and the English built longer and narrower ship type. Instead, the French had developed a larger, sturdier and thus heavier vessels. Because of that, they were also slower. To address this problem, they built hulls with rounded floor timbers and with deep drafts, so that they could slice through water more easily. However, deep drafts were combined with high castles, which generated a significant lurch. Consequently, heavier weights had to be in the lowest deck to stabilize the vessel. Which meant that larger calibre guns should be collocated in lower decks; thus these cannons could not be used with rough sea and they were low efficient even in good sea conditions. And vessels with deep drafts also had two

more problems: they could run aground in shallow waters and harbour approaches were more difficult.
Furthermore, France had heterogeneous shipbuilding traditions depending on the geography. Arsenals placed in the Atlantic coast made narrower hulls, which often needed to be broadened, meanwhile Mediterranean tradition built vessels with a raked bow, which meant that those ships were not adequately supported in rough seas. Since 1670s shipwrights moved from one arsenal to another in order to decrease these differences [17]. François Pomet (17th century) had worked at Toulon until 1669, when he moved to Rochefort; Jean-Pierre Brun (17th century), which probably had studied the English shipbuilding in Britain [19], moved to Brest from Rochefort; the Italian Blaise Pangolo (1650-1714 or 1719) worked at both Brest and Toulon, and the Dutch shipwright Voon (17th century) was at La Havre in 1672 and at Rochefort in 1679 [20]. Nevertheless, the Mediterranean tradition prevailed on the western one, to the point that the Dutch shipwright Rodolphe Gédéon (Unk-1672), who had been working at Toulon's arsenal since 1645 [21], adapted to local styles of construction over time. The Flemish shipwright Hendrik Houvens (Unk-1690) [22] and his son used a Nordic construction technique. But when Hendrick died, his son was replaced by René Levasseur (1667-1727) from Toulon and no one kept using that technique [23]. Besides, northern countries did not seem particularly keen on sharing their shipbuilding secrets with the French. In the 1660s, Colbert tried to draw foreign shipwrights in French arsenals, especially from the Netherlands, in order to let them teach their art to local carpenters. As successfully happened to Pomet, who was trained by Gédéon at Toulon [24]. At the same time, Louis XVI tried to hire the carpenter Anthony Dean (1633-1721), who was one of the best English shipwrights of that time. However, both Colbert and the Sun King received negative answers. the Netherlands granted France six Second Rate vessels built in 1666. These were the *Conquérant*, the *Courtisan*, the *Invincible*, the *Normand*, the *Intrépide* and the *Neptune*. All ships left from Amsterdam to France on 5 May 1667 [25]. English concessions were even minor. The best that Louis XVI could obtain by his cousin Charles II (1630-1685) in the name of the Anglo-French alliance was that Dean built two small yachts for the pleasure of French nobility [26].

In this context, the issue of decoration is no less important. The comparison with unadorned English and especially Dutch vessels pushed admirals and captains to breaking point; they were also supported by the evidence of the foreign fleets' better performance. The naval officer Guillaume d'Almeras, marquis of Mireval (1610-1676) was one of the most fervent opponent of ships decoration. He asked the elimination of galleries in a *memorandum* to Colbert and added a comment on Puget, in which he said that the King would do better to pay Puget to stay outside the arsenal [27]. However, meanwhile England went further by following the Dutch example and abolishing the projecting galleries and decreasing the decoration, many French naval officers did not agree with that position. As testified by Louis de Matharel (1619-1673), the naval intendant of Toulon since 1665, in a letter to Colbert dated June 26, 1671: «*si M. d'Alméras a trouvé les dessins défectueux en quelque chose, la plupart des autres capitaines n'ont pas été de ce sentiment; e il est certain que le sieur Puget donne un tour à ses dessins qu'on ne voit point chez les autres nations*». This difference of opinions prompted France not to follow the radical Dutch choice and led the shipbuilding just to slightly decrease decorations, reducing both costs and weight of vessels [28].

Colbert could not ignore the superiority of foreign fleets and thus, in 1671, he sent one of his councillors to investigate on English and Dutch Navies again; Jean-Louis Girardin de Vauvré (1647-1724) was the naval intendant from 1680 to 1716. He reported the same message of his predecessors, namely that Nordic vessels were lighter and faster and thus better than French ones.
Captains and officers' reaction was that to ignore the critics on French shipbuilding, still convinced to follow the tradition. The strongest reaction came from shipwrights, especially from Brest and Toulon. They could not admit the superiority of another Navy and they took up the cudgels for French shipbuilding. Thus, there were two different ways of thinking between those who supervised and built vessels and those who tested vessels on the sea. On 13th of September 1673, Colbert enacted an Establishment about shipbuilding, which aimed to homologated French vessels with foreign ones. It also provided for a reduction of decorations and the abolition of galleries in the stern. However, given the hostility showed especially by shipwrights, shipyard did not adopt these rules quickly. Moreover, when the Establishment was emanated, the French Navy was mostly composed by newly built vessels. Considering that the average life of vessels was between 15 and 20 years, France had to wait for the last decade of the seventeenth century to have a new generation of vessels built under that Establishment [29]. From 1694 to 95, the financial strait and a higher use of privateering stimulated the application of smaller vessels, boarding between 20 and 60 cannons. Indeed, these kinds of ships had lower construction costs and were more serviceable than First Rates. Subsequently, from the 1720s onwards the French Navy started a renewal of its fleet involving longer hull shapes, which were thus faster and more suitable for wars in the ocean [30]. Whilst France started to reduce decoration at the end of the seventeenth century, English shipwrights had begun earlier to follow the Dutch way, as shown by the *Royal Sovereign* built from 1697 by Fisher Hardling (Unk-1705) at the Woolwich dockyard and launched in 1701 [31]. Actually, the sides of this vessel were almost unadorned, such as the bow. However, the stern part was still high decorated with elaborately carved ornaments; but the expensive realization led the Admiralty decided to limit carving and sculpture presence onboard.

The political condition has to be taken into account to understand the reason of these different behaviours. The French absolute monarchy was quite different from the English policy, because, since XVII century, England has become a constitutional monarchy. Thus, the Sovereign did not have full control of the national treasury; consequently, the proposal to build a vessel would not have been accepted if cost of excessive decoration aimed at glorifying the monarchs had been high [32]. On the contrary, Colbert managed to afford high cost of shipbuilding thanks to his autonomy in economic management. Indeed, he held many political offices; he was superintendent *des bâtiments*, superintendent *des arts et des manufactures* and he also managed the *contrôle général des finances*. Therefore, he could work synergistically in different fields [33]. In addition, Louis XVI was not expert in naval affairs and even if he attended the *Conseil de Marine*'s meetings, he did not have technical skills to take decision about that; the King thought of vessels mainly as a means of political propaganda [34]. Unlike his French counterpart, Charles II was aware of the importance of naval power and therefore paid particular attention to the naval affairs after the restoration. James II (1633-1701) reigned after him; the new sovereign was a skilled admiral and he engaged in Navy development too, through the eyes of an expert of naval affairs. Later, William of Orange (1650-1702) became king from 1689; he proved to be a brilliant strategist also in naval warfare, so much so that in 1690 he asked the Parliament to allocate annual funds in order to the expansion of the fleet [35].

The divergence with the western European shipbuilding led to a negative opinion on French fleet in foreigners' minds. Edmund Dummer (1651-1713), for instance, was a English shipwright who visited Toulon in 1638; having observed the fleet which was there stationed, he reported: *«A good fleet, but in appearance ill built, or through some weakness in long living have generally put their wales a little straight in the midships»* and adding about the *Royal Louis*: *«a great ship and glorious in her first carving, no doubt; but to my judgment, not of good proportion, nor good workmanship, her figure under water I know not, nor is that above to be admired»* [36]. But in King Sun's vessels, the 'first carving' had the same value of 'good proportion'.

3. MASTERPIECES AFLOAT, THE *ROYAL LOUIS* AND ITS CONTEMPORARIES

The *Royal Louis* [Figure 2] is a perfect example of the masterpieces afloat launched in the France of seventeenth century. From 1669 to 1692, she had been the flagship of Mediterranean French fleet based at Toulon. Her construction lasted from 1666 to 1668 and was directed by the aforementioned Rodolphe Gédéon. At the time this vessel was without equal throughout France. She was one of the first *full three-decker* ships; that means, she had three full decks totally armed with guns, meanwhile, usually, the so called "three-decker" vessels had two full decks and the upper one was actually an half-deck. The *Paris* (1669) was the last vessel of that kind built by the French Navy [37].

The *Royal Louis* was the biggest French vessel of her day; she measured 2.400 tons and 163 feet (about 49 meters), with a beam of 44 feet (about 13 meters), and a height over the waterline of 48 feet (about 14.5 meters). Originally thought for boarding 800 men plus 9 officers, later she embarked about 850 men. Her armament exceed other vessels too, since she had been designed for 110 guns located in decks, forecastle and quarterdeck. And even if the effective number of guns was reduced at 104 bronze pieces of artillery, it still exceed the average, considering that First Rates generally boarded from 70 to 100 guns. For these reasons, she was rated as a *vaisseau du premier rang extraordinaire* along with the *Soleil Royal* (1670), which was her Atlantic counterpart. Thus, they were considered as extra-rate vessels and that perception had been strengthened through an Establishment enacted on 4 July 1670, which imposed they had to be the only two vessels with forecastles [38]. The importance of the *Royal Louis* was immediately claimed, and the very fact that she was called after the King credited to her a certain supremacy on other vessels. The high regard for the *Royal Louis* was also confirmed by the motto wrote in gold letters at the base of the mizzenmast: *«je suis l'unique dessus l'onde, / et mon Roy l'est dedans le monde»*.

Figure 2: Drawing of the *Royal Louis*. By: Maître Rodolphe. In: Hayet. *Description du vaisseau le «*Royal Louis*»*. Marseille: Charles Brebion, 1677. Source: gallica.bnf.fr / Bibliothèque nationale de France.

In 1677, the superintendent of Toulon Hayet described the *Royal Louis* with these words: *«On peut dire que jamais aucun Navire n'a esté si enrichy de peinture & de sculpture que cet incomparable vaisseau»* [39]. Considering the importance of this vessels, it is not surprising that also the decorations were at the highest level. Hayet wrote this comment in a report dedicated to Pierre Arnoul, who had been *intendant général de la Marine de Levant* from 1675 to 1678 [40], describing in detail the flagship. According to this description, the interior of the vessel was richly ornamented with gold, and refined paintings were hung on the walls. The

Chambre des Volontaires (or *Chambre du Conseil*) and the *Corps de garde* [41] were two of the most decorated areas, both placed on the middle deck. The *Chambre des Volontaires* contained golden mouldings enriching the environment and surrounding three golden frames. Inside, there were displayed the *Armes du Roy* and the coat of arms of César de Bourbon duke of Vendôme (1594-1665), who became the *Grand-maître de la navigation* after Anne of Austria until 1665, and his son François de Bourbon-Vendôme duke of Beaufort (1616-1669), who took the same office after his father until the 1669; that is because the *Royal Louis* should have been served as the flagship of the latter. The floor was made by precious materials as olive and ebony wood and ivory [42]. Inside the room there were also two big paintings depicting mythological scenes about "Apollo and Python" from Ovid's Metamorphoses. One portrayed Cupid stringing his bow towards Apollo and the other one showed Apollo chasing the nymph Daphne. The *Corps de garde* was below the stern castle. King's initials found their place over a golden globe above the gateway. Inside the room, a big table was in the middle; its legs were eight columns with capitals all in red jasper. On the wall there were four gunports decorated by *cartouches* [see Appendix 1] and between them landscape and seascapes paintings were hung in golden frames. These two rooms had light blue ceilings on which decorations in gold and *grisaille* [see Appendix 2] depicted royal symbols as lilies, suns, crowns and royal initials. Also the naval officers' rooms were in the stern part and were similarly decorated. On the walls there were paintings with mythological scenes or Royal symbols and ceilings had again golden decorations on a light blue background. Finally, on the back of the upper deck, there was the *dunette* [43]. This was the area reserved to the rooms of high-ranking officers (such as the *maître* or the *pilote*). In the *Royal Louis*, like in the biggest warships, the *dunette* was divided into two separated areas. The first had a hallway with four doors, which were embellished with colourful floral decorations. Inside, these rooms had the same splendour and wealth of the other areas. Everything was enriched by floral decorations, which surrounded the monarchy's symbols; paintings showing seascapes and battleship scenes were enclosed in golden frames. The second area of the *dunette* was in a more elevated position and contained eight smaller rooms decorated as the other parts. Nothing on the *Royal Louis* was left totally unadorned. To the point that even the powder magazine was decorated by drawing of *fleurs-de-Lys* enriched by golden thread. And so it was for the staircase leading to the upper deck, whose balustrades were made by painted panels.

The exterior was no exception. The bow was fully covered by statues. A representation of The Fame was surrounding by sea creatures as mermaids and mermen. Golden floral decorations run throughout the sides of the *Royal Louis* embellishing the balustrades and the gunports. The stern was the most sumptuous area; it was the stage on which artists could display their skills. Here, mermaids, mermen and seahorses seemed nestled in the galleries for a spell. In the upper side there was the greatest sculpted group; a wooden statue of Louis XIV was sitting on a throne of justice holding a palm leaf as symbol of peace and a laurel branch for the victory. At the feet of the king there were two Turkish slaves, who represented the submission of unbelievers to the Christendom. On the sides of the sovereign there were Thetis and Neptune offering him respectively the goods of the Earth and of the Sea.

The craftsmanship required had to be able to realized artworks capable of showing the French monarchy magnificence. Skilled artist were thus called to work on the *Royal Louis*. The painter Jean-Baptiste de la Rose (1612-1687) was in charge of controlling the drawings, and François Girardon (1628-1715) [44], who was an important sculptor of his day, had to oversee the work and also made himself the statues of greatest importance. He was supported by Pierre Puget, who was particularly appreciated for his sculptures. There were also five teams composed of lesser-known sculptors as Nicolas Levray and Rombaud Languenu (1638-1718), who had to make all the decorations. Furthermore, the drawings for the major decorations had been realized by the Versailles' painter Charles Le Brun (1619-1690).

In Toulon's arsenal, there were many vessels under construction while the *Royal Louis* was being built. The *Monarque* (80 guns), which was built from 1666 to 1669, the *Royal Dauphin* (100 guns), built between 1666 and 1671, the *Paris* (80 guns) and the *Ile-de-France* (80 guns) [45], both built from 1667 to 1671, were the biggest ones. Hence, they were also richly decorated and their realization required a large number of artists. The increasing presence of artists inside French dockyards led to tense situations between them and other craftsmen. The superintendent of the artwork of Toulon Louis Le Roux d'Infreville (1642-1712) complained to Colbert about the artists behaviour in a letter of 21 April 1668 saying that «(...) *je feray bien mon possible pour les tenir en leur devoir, mais il est absolument nécessaire d'avoir un commandant comme le sieur Girardon ou une personne de sa suffisance pour conduire un sy bel ouvrage et assujettir les gens de ce mestier* (...)» [46]. Even the judgment about the artwork was not always positive, as showed in a letter of 1670 from the superintendent Matharel to Colbert: «*les sculpteurs de la marine s'attachaient plus aux règles de leur art et à la démangeaison de faire de belles figures qu'au besoing, commodité et service du navire*» [47].

However, real *ateliers* were created inside the Toulon's arsenal, which turned out to be appealing to many others artists. Some of them were foreigners, especially coming from Italy and Flanders. This caused a migration phenomenon and «*entre 1670 et 1680 plus de quatre-vingts sculpteurs et cinquante-cinq peintres se côtoyèrent ainsi à l'arsenal de Toulon, faisant de la ville l'une des plus peuplées de France en nombre de peintres et sculpteurs par habitant*» [48]. The other important French dockyards were experiencing a similar situation;

the monarchy hired artists in order to secure the exclusive of their work and that led to a sort of artistic patronage [49]. The importance of artworks was quickly increasing. So much that, in 1672, Girardon sent to the dockyards of Brest, Marseille and Toulon, 948 books of «*modèles de plâtre, dessins et estampes*» [50], which were useful for a refinement of the artistic technique. The advantage of training artists right in the workplace was becoming evident and it led to the creation of dedicated structures. The *Académie de Marseille* was founded in 1752 and two years later a training programme started for artists bound to work in French arsenals. The other two schools for artistic training were located in Paris; they were the *École des Beaux-Arts*, which was founded in 1648 at the behest of per Cardinal Mazarin (1602-1661) and the *Académie royale d'architecture*, which was established in 1671 thanks to Colbert.

The choice of adorn vessels so much, led inevitably to high shipbuilding costs. In a letter to Colbert dated 4 November 1667, d'Infreville reported the estimated costs for the realization of «*sculptures, dorures et peintures des trois desseins par eux dressez de la poupe*» of the *Royal Louis*. For De la Rose's project, the cost was about 37.060 *livre tournois* [see Appendix 3]. 21.300 *livre* had been estimated for the realization of sculptures and 15.760 *livre* «pour la despense de l'or, peinture et ouvriers qu'il faut pour decorer et peindre» the sculptures. The cost estimated for Languenu's labour was of 25.850 *livre*, and 23.800 *livre* was the necessary for paying the Levray's team. The total amounted to almost 89.000 *livre* [51]. Ten years later, considering the change in the currency's value, Hayet wrote in his report that the cost of the *Royal Louis* was of 65.800 *livres*, whose 20.000 *livres* for the decoration and 27.000 *livres* for the carpentry [52]. Thus, artwork cost a little less than carpentry work. Hence, monarchy was willing to spend the same amount of money for ornamentation and structural parts. That means, aesthetics and seaworthiness had the same level of importance.

3.1 THE FLAGSHIPS' INACTIVITY

The *Royal Louis* was launched on 1st of February 1668; she entered service in the August 1669. Despite her great fame, her life had not been characterized by great enterprises. On the contrary, it was almost without military events. She was built to be the flagship of the duke of Beaufort, illegitimate grandson of King Henry IV of France (1553-1610) and cousin of Louis XIV. She should have left for a Mediterranean campaign immediately after her launch, instead she remained anchored at Toulon's port to further works. On 11 May 1677 she left for her first and last military campaign, leading a naval squadron under Admiral Abraham Duquesne marquis of Bouchet (c. 1610-1688), headed for Messina (Sicily) to support rebels against Spanish dominance of the island. After that, no other war actions are reported for the next 14 years, until she was disarmed in 1691. In 1692 a homonymous vessel was lunched and, therefore, the old *Royal Louis* was renamed *Royal Louis Vieux*. After almost 20 years of inactivity, she was finally broken up in 1697 [53].

The *Soleil Royal* [Figure 3], which was the western fleet's flagship, had a similar story. She was built at Brest by the shipwright Laurent Hubac (1607-1682) at the same time of the *Royal Louis*. Despite she was launched in 1671, she participated to her first campaign only in 1690, during a mission aiming to land on the English coast. She was opposed by the English flagship *Sovereing of the Seas* during the battle of La Hougue. The *Soleil Royal* was overpowered by the enemy and had to withdraw. After that, in 1692, she sank off the coast of Cherbourg in the battle of Barfleur [54].

Figure 3: Poupe du Soleil Royal. 1669. By : Jean Bérain. Musée du Louvre, D.A.G. Paris. Credit : © Musée du Louvre, Dist. RMN-Grand Palais / Martine Beck-Coppola

The correspondence of the marquise Marie de Rabutin-Chantal (1626-1696), known as Madame de Sévigné, with her daughter Françoise-Marguerite de Sévigné, countess of Grignan (1646-1705), has became famous for the many indiscretions about the coeval social life reported inside. In a letter of 1672 there is an interesting news; indeed, she wrote: «(...) Rien n'est plus romanesque que vos fêtes sur la mer, et vos festins dans le *Royal-Louis*, ce vaisseau d'une si grande réputation» [55]. This suggests us that French aristocracy could find entertainment for its social life on board these wonders of the Sea, not used for naval combats.

4. CONCLUSIONS

The common destiny of the two French flagships can perhaps make them appear as a untapped potential if put in relation with the economic effort required and the great deal of resources including time spent for their construction. But the truth is that these enormous and astonishing *three-deckers* built in seventeenth century European Navy had high cost of maintenance. The huge number of men needed to sail them was of eight hundred men for the biggest vessels, and it substantially arose the operating costs. Moreover, also the repair costs of any damage to the wooden and golden decorations could be exorbitant. The cumbersome decoration of the topside adversely affected aerodynamics, making these vessels less manoeuvrable and slower. Sir Walter Raleigh (1552 or 1554-1618), who was one of the s*ea dogs* of Elisabeth I (1533-1603), said on the matter: «*We find by experience that the greatest ships are least serviceable, go very deep to water, and of marvellous charge and fearful cumber. ... besides they are less nimble, less mainable, and very seldom employed*» [56].
What was the purpose and what the advantage of building such impressive vessels, despite the high costs, if they were not useful for naval operation? It is clear that their value did not lie in their practicality but it was related to the symbolic character of power that a Navy took having that kind of ships in its fleet. For instance, when Charles I ordered the construction of the *Sovereign of the Seas*, he was pushed by a sort of competition with other sovereigns. And it was especially a reaction to the launch of the *Couronne*. First Rates had an *institutional* role that went beyond the mere use in military actions, because they were mainly used for purpose of representation. Flagships were symbols of power and an ostentatious display of strength. In this sense, they were *cultural icons* with a specific identity, which tried to find its continuity by reusing the same names in the following flagships [57]. And that is the reason of the success of the name "*Royal-Louis*", which was used for the eighth time in the homonymous *three-decker* launched in 1811. And equally, many vessels were named *Sovereign* in the English Navy. They were the Navies' pride which was shown as a warning to the other European powers.

And hence, *appearance*. The same appearance that is suggested by architecture, fashion and generally by the Baroque style. *Appearance*. Today, this term has taken a negative meaning, as a synonym of superficiality. Nevertheless, the ostentation of power still have a place in the World and in the seas. It is a consequence of the constant arms race that has always put States in competition with each other. It has surely changed the means, but maybe not the essence.

5. APPENDIXES

1. Cartouches: bas-relief in floral theme. For examples of cartouches in Versailles style: *Ornamental borders scrolls and cartouches in historic decorative styles*, Syracuse Ornamental Company, New York: Dover Publications, 1923; especially plate 92.

2. « Grisaille: A painting done entirely in shades of grey or another neutral greyish colour. [...] » In: CHILVERS, I., *The Oxford Dictionary of Art*, 3rd edition, Oxford University Press, 2004; p. 314.

3. The *livres* were a coin created from a piece of metal. Giving the weight of that piece, there was a specific number of coins that could be mint. The value of the coins was thus depending on the weight/number of coins ratio. Frequently, weights and measures did not coincide between different places and thus there were *livres* with different values. To distinguish them, people added the places of origin to the common prefix. Among those coins, the most used were the *livres parisis* (lp) and the *livre tournois* (ŧt). The lp/lt ratio was 1,25. Since 1667 the *livres parisis* have been officially abolished and being disappeared every ambiguity, the coin could be called just *livre*.

6. ACKNOWLEDGEMENTS

This publication was written in "Art of building and aesthetics in naval iconography from the Middle Ages to the Modern Age" research area and it was financed with Department of Architecture and Design 'PRA 2019' funds of the University of Genoa

7. REFERENCES

1. GLETE, J., *Warfare at Sea, 1500-1650: Maritime Conflicts and the Transformation of Europe*, London: Routledge, 2000.

2. FILION, M., 'La crise de la Marine française, d'après le Mémoire de Maurepas de 1745 sur la marine et le commerce', *Revue d'histoire de l'Amérique française*, Vol. 21, n° 2, septembre 1967; p.231.

3. CIPOLLA, C. M., *Vele e cannoni*, Bologna: il Mulino, 2019; p. 231.

4. ROBERTSON, F. L., *The evolution of naval armament*, London: Constable & Company LTD, 1921; p.21.

5. CORBETT, J., 'Fighting Instructions, 1530-1816', *Publications of the Navy Records Society, Vol. XXIX*, 1905; pp. 92-93.

6. DULL, J. R., *The Age of the Ship of the Line. The British and French Navies, 1650-1815*, Lincoln and London: University of Nebraska Press, 2009.

7. WINFIELD, R., *British Warships in the Age of Sail 1603-1714: Design, Construction, Careers and Fates*, Barnsley: Seaforth Publishing, 2010; pp.5-6.

8. BRUIJN, J. R., *The Dutch Navy of the Seventeenth and Eighteenth Centuries*, Oxford: University Press, 2017; pp. 17-18.

9. PAINE, L. P., *Warships of the World to 1900*, Boston and New York: Houghton Mifflin Harcourt, 2000; p. 43.

10. WINFIELD, R., *First Rate: The Greatest Warships in the Age of Sail*, Annapolis: Naval Institute Press, 2010; pp.13-17.

11. DESSERT, D., *La Royale, vaisseaux et marins du Roi-Soleil*, La Flèche: Fayard, 1996; p. 20.

12. BRUN, V. F., *Notice sur la sculpture navale, et chronologie des maîtres sculpteurs et peintres du port de Toulon*, Toulon: D'E. Aurel, 1861; pp. 13-15.

13. CHACK, P., *Marins à la bataille*, Vol. 1. Paris, Le Gerfaut: 2001; p. 169.

14. TOURNEUX, M. 'Table générale des documents contenus dans les archives de l'art francais et leurs annexes (1851-1896)', Paris, *Charavay Fréres*, 1897; p.116.

15. LAGRANGE, L., *Pierre Puget: peintre, sculpteur, architecte, décorateur de vaisseaux*, Paris: Librairie Académique, 1868; p. 15.

16. WINFIELD, R.; STEPHEN R., *French warships in the age of sail 1626-1786: design, construction, careers and fates*, Barnsley: Seaforth publishing, 2017; pp.4-5.

17. DESSERT, D., *La Royale, vaisseaux et marins du Roi-Soleil*, La Flèche: Fayard, 1996; pp. 130-138.

18. CHALINE, O., *La mere et la France. Quand les Bourbons voulaient dominer les océans*, Lonrai: Flammarion, 2016; p.241-2.

19. ANTHIAUME, A., *Le Navire. Sa construction en France et principalement chez les Normands*, Paris: Librairie Ernest Dumont, 1922; p.320-21.

20. ACERRA, M., 'Les constructeurs de la marine (XVII°-XVIII° siecle)', *Revue historique*, by MM. G. Monod e G. Fagniez, Paris: Presses universitaires de France, 1985; p. 284.

21. BOUDRIOT, J., *Le vaisseau Trois-Ponts du Chevalier de Tourville*, Paris: Collection Archeologie Navale Française, 1998; p. 10.

22. ANTHIAUME, A., *Le Navire. Sa construction en France et principalement chez les Normands*, Paris: Librairie Ernest Dumont, 1922; p.319-20.

23. DESSERT, D., *La Royale, vaisseaux et marins du Roi-Soleil*, La Flèche: Fayard, 1996; p. 84.

24. ACERRA, M., 'Les constructeurs de la marine (XVII°-XVIII° siecle)', *Revue historique*, by MM. G. Monod e G. Fagniez, Paris: Presses universitaires de France, 1985; p. 285.

25. WINFIELD, R.; STEPHEN R., *French warships in the age of sail 1626-1786: design, construction, careers and fates*, Barnsley: Seaforth publishing, 2017; pp. 81- 2.

26. DESSERT, D., *La Royale, vaisseaux et marins du Roi-Soleil*, La Flèche: Fayard, 1996; p. 81-5.

27. BRUN, V. F., *Notice sur la sculpture navale, et chronologie des maîtres sculpteurs et peintres du port de Toulon*, Toulon: D'E. Aurel, 1861; p. 20.

28. BRUN, V. F., *Notice sur la sculpture navale, et chronologie des maîtres sculpteurs et peintres du port de Toulon*, Toulon: D'E. Aurel, 1861; p. 21.

29. DESSERT, D., *La Royale, vaisseaux et marins du Roi-Soleil*, La Flèche: Fayard, 1996; pp. 144-6.

30. CHALINE, O., *La mere et la France. Quand les Bourbons voulaient dominer les océans*, Lonrai: Flammarion, 2016; p.240.

31. WINFIELD, R., *First Rate: The Greatest Warships in the Age of Sail*, Annapolis: Naval Institute Press, 2010; p.45.

32. WINFIELD, R., *British Warships in the Age of Sail 1603-1714: Design, Construction, Careers and Fates*, Barnsley: Seaforth Publishing, 2010; pp.5-6.

33. LA RONCIÈRE, C. De, *Histoire de la marine française*, Vol. 5. Paris: E. Plon, Nourrit. 1899; p. 337.

34. DESSERT, D., *La Royale, vaisseaux et marins du Roi-Soleil*, La Flèche: Fayard, 1996; pp. 19-20.

35. FINCHAM, J., *A History of Naval Architecture*, London: Whittaker & Co., 1851; pp. 60-6.

36. FOX, C., 'The Ingenious Mr Dummer: Rationalizing the Royal Navy in Late Seventeenth-Century England', *The Electronic British Library Journal*, 2007, article 10; p. 17.

37. WINFIELD, R.; STEPHEN R., *French warships in the age of sail 1626-1786: design, construction, careers and fates*, Barnsley: Seaforth publishing, 2017; pp. 55-6.

38. LEMINEUR, J.C., 'Les vaisseaux du Roi Soleil: étude historique', Nizza, Omega « *Collection Archéologie navale française* », 1996; p. 41.

39. HAYET, *Description du vaisseau le «Royal Louis»,* Marseille: *Charles Brebion*, 1677; p. 25.

40. DESSERT, D., *La Royale, vaisseaux et marins du Roi-Soleil*, La Flèche: Fayard, 1996; p. 40.

41. AUBIN, N., *Dictionaire de Marine contenant le termes de la Navigation et de Architecture navale*, Amsterdam : 1702; pp. 187, 279.

42. HAYET, *Description du vaisseau le «Royal Louis»,* Marseille: Charles Brebion, 1677; p. 26.

43. AUBIN, N., *Dictionaire de Marine contenant le termes de la Navigation et de Architecture navale*, Amsterdam: 1702; p. 336.

44. LAGRANGE, L., 'Toro', *Gazette des beaux-art. Courrier Européen de l'art et de la curiosité*, vol. 25, Paris, Impr. de J. Claye, 1868; pp. 347-352.

45. LAGRANGE, L., 'Toro', *Gazette des beaux-art. Courrier Européen de l'art et de la curiosité*, vol. 25, Paris, Impr. de J. Claye, 1868; pp. 347-352.

46. LAGRANGE, L., 'Pierre Puget : peintre, sculpteur, architecte, décorateur de vaisseaux', Paris, Didier et Cie, 1868; p. 113.

47. COUFFON, R., *La Sculpture au Port de Breste aux XVII et XVIII Siècle*, Saint-Brieuc: Les Presses Bretonnes, 1951; pp.1-2.

48. THÉRON, M., 'Les ateliers de peinture et de sculpture des arsenaux en Provence en marge de l'Académie de peinture et de sculpture de Marseille' *Rives méditerranéennes*, 2018/1 (n. 56); p. 150.

49. LACROIX-LINTNER D., 'Marine, beaux-arts et mécénat au XVIIe siècle en France', *124-Sorbonne. Carnet de l'École Doctorale d'Histoire de l'Art et Archéologie*, 2016; pp. 3-8.

50. THÉRON, M., *L'ornementation sculptée et peinte des vaisseaux du Roi*, Tome 1. Paris: Université de la Sorbonne, 2003; pp. 81-83.

51. DE MONTAIGLON, M. A., *Archives de l'art français. Recueil de documents inédits relatifs à l'histoire des arts en France*, Paris: J.-B. Dumoulin 1855-6; pp. 233-5.

52. HAYET, *Description du vaisseau le «Royal Louis»*, Marseille: Charles Brebion, 1677; p. 11.

53. WINFIELD, R.; STEPHEN R., *French warships in the age of sail 1626-1786: design, construction, careers and fates*, Barnsley: Seaforth publishing, 2017; p. 57.

54. DESSERT, D., *La Royale, vaisseaux et marins du Roi-Soleil*, La Flèche: Fayard, 1996; p. 327.

55. RABUTIN-CHANTAL, M. de, *Lettres de Madame de Sévigné, de sa famille et de ses amis*, Tome III, Paris: Lahure Hachette et Cie, 1862; p. 78.

56. OLDYS, W.; T. BIRCH, *The works of sir Walter Ralegh*, Volume 8, Oxford: University press, 1829; p. 337.

57. WINFIELD, R., *British Warships in the Age of Sail 1603-1714: Design, Construction, Careers and Fates*, Barnsley: Seaforth Publishing; 2010.

9. AUTHORS BIOGRAPHY

Claudia Tacchella is a PhD candidate in the field of History of Shipbuilding at the Polytechnic School of the University of Genoa (Italy).

THE ENTRAILS OF A SHIP: A NAUTICAL POST-MORTEM

D. A. Walker, Maritime Museum of the Atlantic, Canada

SUMMARY

The largest man-made explosion prior to the nuclear bomb occurred during World War I in Halifax Harbour, Nova Scotia, Canada when a ship loaded with war munitions collided with another ship and detonated almost spontaneously. The author examines the retrieved shrapnel from the completely disintegrated ship to find clues to the extent and the effects of the explosion, and the distribution of the detritus. Many inquiries, studies and books have concentrated on the effects of this explosion on the shattered city, the population, buildings, emergency preparations, &c., but until now none have looked exhaustively at the ship, the floating bomb and its dispersed detritus.

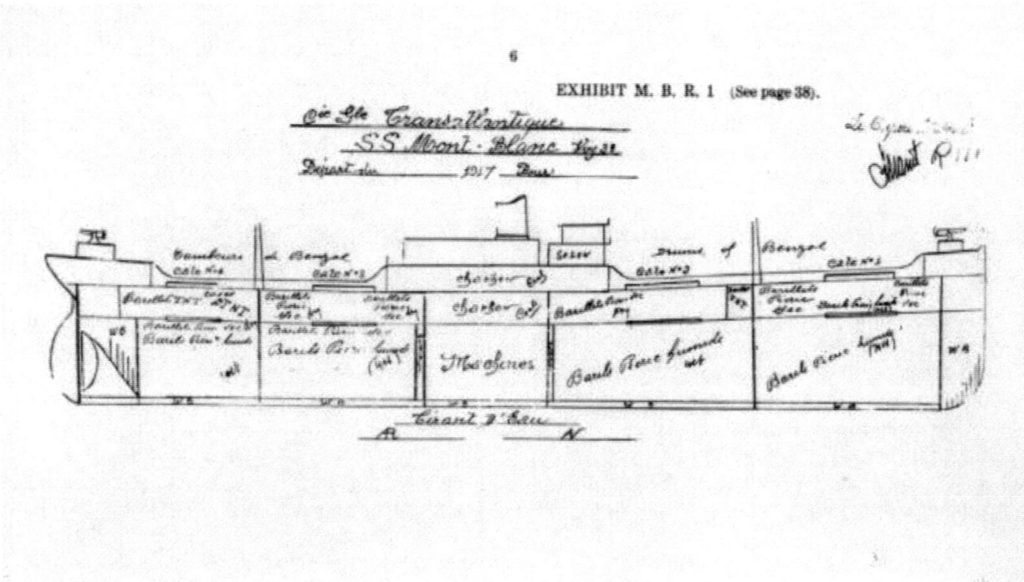

Figure 1: Cargo Plan prepared for the official inquiry into the collision.

This will not be a conventional medical autopsy as there are insufficient body parts to examine and we already know how the patient died. It is more akin to archaeology where fragmentary chards are analysed and a small fragment of pottery may point to a previously known technology and explain the entire context. The scalpel has been exchanged in favour of historic records of contemporaneous shipbuilding methods.

SS ***Mont-Blanc*** was an utterly bog-standard, three-island tramp steamer launched in 1899 from Sir Raylton Dixon's Shipyard in Middlesbrough. She was merely one among scores of this type of general cargo ship gliding down slipways monthly from dozens of British shipyards in the latter quarter of the nineteenth century. Most were bought by British cargo fleet owners supplying the world with British manufactured products and returning with raw materials. But many were also built for foreign owners operating cargo fleets in competition. ***Mont-Blanc*** was launched in 1899 and exported to the French company Société Générale de Transports Maritime. An interesting aside regarding the launching formalities of these ships, at least the foreign owned ones – no fuss was made. They merely entered service like microwaves coming from Bosch, they were produced and shipped. No mention in the press, not even the local press in Middlesbrough noted ***Mont-Blanc***'s entry into service. Entries in official records: national registries, insurance documents, crew lists and Lloyds Register, &c., sufficed.

The steamer went on to lead what must be a near typical life of a cargo vessel at the beginning of the 20th century. She changed port of registry and owner's titles but always flew the French tricolour. She ran into maritime difficulties occasionally and at one point entered the Halifax, NS shipyard for repairs. We come to 1917 when ***Mont-Blanc*** then under the ownership of Compagnie Générale Transatlantique and under wartime governmental charter, loaded a cargo of war munitions in Brooklyn destined for Bordeaux.

Orders directed the ship to sail for Halifax, NS, Canada to join a slow convoy destined for Europe. The German U-Boat threat had made convoys necessary as slow ships sailing alone were extremely vulnerable – ***Mont-Blanc*** was a 9.5 knot flyer.

The cargo carried by ***Mont-Blanc*** was roughly as follows: (In tonnes) Benzole, 246; Gun-cotton, 62: Picric Acid, 2,367: TNT, 250 and just to round out the load some 300 rounds of 90 and 95mm cannon ammunition. Some of the latter perhaps was supplied for the ship's two defensive guns. A more volatile cargo for that period is hard to imagine. This vessel was a mobile, floating bomb of enormous proportions.

On the evening of the 5th December, ***Mont-Blanc*** arrived off the Port of Halifax and picked up her pilot, but when the ship approached the harbour entrance the guardship refused entry as the submarine net had been raised for the night. The steamer was brought to anchor off McNabb's Island and the pilot stayed aboard for the night. Similarly in Bedford Basin at the northern end of the harbour SS ***Imo*** had been refused exit passage on the evening of the 5th. She was a neutral, Norwegian registered steamer bound for New York to pick up a cargo of materials destined for the relief of the beleaguered Belgian population. Her refuelling coal barge had arrived too late to finish loading in time for a departure before the submarine nets were raised.

The following morning at about 0845 the two vessels got underway in opposite directions. ***Imo*** to leave Bedford Basin destined for New York while ***Mont-Blanc*** left her anchorage and proceeded up-harbour planning to re-anchor to wait orders in Bedford Basin. Through a series of contradictory pilot's orders and misinterpreted audio signals the two vessels collided at the narrowest part of the harbour – appropriately known as The Narrows. ***Imo*** *ran* into ***Mont-Blanc*** her high bows knifing into the starboard side of the laden-to-capacity number one hold. The ship's drifted apart and ***Mont-Blanc***'s crew took to their lifeboats in great haste, they knew all about their dreadful cargo. The unmanned ship drifted away to end up almost alongside a wharf on the Halifax city side of the harbour.

Upon impact of the ships sparks flew and the deck cargo of benzene was broached and ignited. The flaming benzene flowed down into the ruptured hold and at 0904 on the 5th of December, 1917 the cargo exploded in a rapid almost instantaneous cataclysmic eruption. A tsunami tore through the harbour and ***Mont-Blanc*** disappeared. But not entirely, as the mass of iron at her heart stayed put, the triple expansion steam engine & two boilers remained in situ. The force of the explosion dissipated in every conceivable direction a 360 degree circle in the horizontal plane and vertically something over 180 degrees, the downward force partially resisted by the shallow surrounding water and immediately thereafter the seabed. Many survivors claimed that the bottom of the harbour was seen and it is not unreasonable the explosive force would blow the water completely away in the periphery of the disintegrating ship or even vaporise the sea to steam.

Figure 2: Part of the sternpost forging with rudder pintle.

The rest, all the torn apart hull shell, cargo and superstructure flew in every conceivable direction. Lethal projectiles, solid shells tearing apart anything in their path, one such projectile completely removed one of HMCS Niobe's funnels. Those which rose higher returned as lethal white hot bombs onto the city; most dropped back into the surrounding sea to disappear and corrode, gone forever; much would drop to earth round the shattered city to be cleared along with the accumulated dross into piles of detritus which were hurriedly removed to vacant land so life cold return and continue. The pieces which fell on wooden city buildings created fires, damaged stouter structures, vehicles and directly killed many people. One killed a member of the fleeing ***Mont-Blanc*** seaman the only crew member to be lost in the tragedy.

These tragic objects lay everywhere, mute testimony to the sheer violence of the event. They became impediments to movements of the populace or emergency vehicles. When they cooled the smaller bits became objects of interest and macabre souvenirs. The collection of these objects has never stopped, it continues today. Just recently a piece came to light as an older building was being

Figure 3: Extremely battered tip of a swing arm davit; Similarly battered links of a small chain run; A typical example of a small object which has been mounted prior to donating to the museum.

razed. The repository for those unwanted pieces is The Maritime Museum of the Atlantic in Halifax, NS.

Figure 4: How rivets were affected by the explosion.

An obvious question is how does one know these diverse objects are part of ***Mont-Blanc***? One unique characteristic which almost all share is their battered condition. The surface of the steel is bruised as if hammered continuously by a series of massive ballpeen hammers. Their edges are ragged, torn away violently from adjacent pieces of steel whether plate, rolled sections or castings. Rivets perhaps remain, securing a connection, or are rent out from a distorted hole. But the most impressive characteristic is the shapes into which the steel has been distorted. Whatever the thickness of the plate or section it is bent, twisted, distorted and misshapen into a Kafkaesque frenzy of steel, a metallic cacophony of convoluted modern art. Heavy castings or forgings showed less surface damage but indicated the strength of the explosion by the fractured edges they display where they were severed away from the whole.

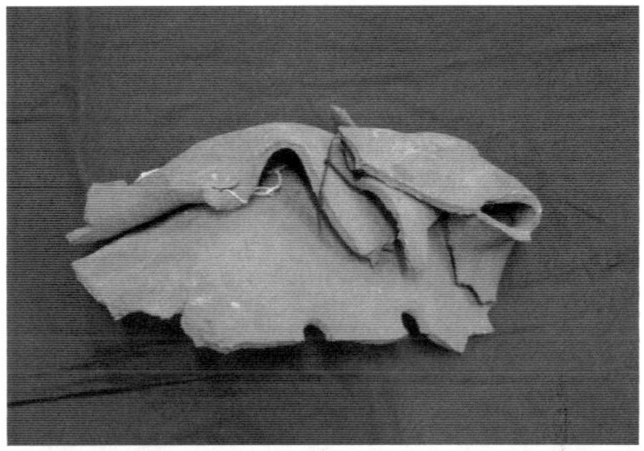

Figure 5: Typical example of a piece of bulkhead (?) plating.

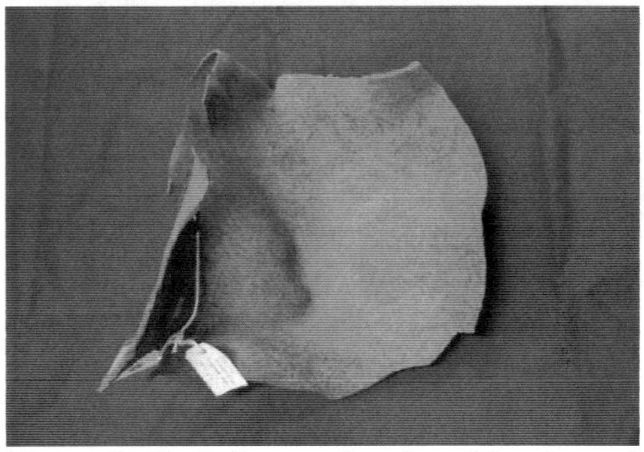

Figure 6: Typical example of a thin piece of plating.

The MMA collection was catalogued and photographs were added to the accession sheets to aid identification of each. It has been a mammoth task, but to use an actual example: it is a small piece of splintered bulb angle still rived to a fractured piece of plate. We can size the bulb angle bar and also the plate thickness and perhaps the rivet dimensions but then where are you? Reference to Lloyds Rules for the early part of the 20th century suggests this could be a deck or a bulkhead plate, we know the tonnage of the ***Monte-Blanc*** therefore can interpret the rules for the sizes but . . . Of course deck plates differ in thickness depending on their location fore and aft or even transversely. The complications of location offer too many choices. On the other hand a shattered forging with an attached pintle is easier to identify or part of a stem-post with tap rivets (again the associated plating was torn away).

Three pieces of organic matter from ***Mont-Blanc*** have survived all are wood, two are wooden name boards normally attached to lifeboats. We know that two lifeboats were lowered and rowed ashore to Dartmouth across the harbour from the epicentre of the explosion, but the boats must have been destroyed by the huge wave created, the boards were found washed ashore nearby. Interestingly one board has been reversed and re-carved changing the Port of Register from Rouen to Marseille. The third piece of wood cannot be definitively associated with ***Mont-Blanc***, it is a segment of a lignum vitae propeller shaft stern bearing. It was found at the low tide line near where the ship disappeared and is (more or less) the correct size, no marks exist with which to connect the two. The association is very positive but of course the word *probably* has to be inserted.

Figure 7: Wooden Port of Registry nameboards from a ***Mont-Blanc*** lifeboat.

The description of some of the foregoing jetsam, the self-distributed detritus which was a ship gives a vivid illustration of the severity of the explosion. Most of ***Mont-Blanc*** lies where the countless pieces fell to earth or water to lie forever undiscovered. However some of the retrieved objects have suggested a theory in where they lay. Appendages to the operating vessel, her anchors, guns, davits, &c. have been blown the furthest, regardless of their size, indicating that the blast simply tore them loose and cast them afar. On the other hand structural items from keel plating to fore and aft end castings to plating-framing connections have stayed closer to ground zero. The energy of the explosion, severe as it was could have been partially exhausted in physically tearing the structure asunder prior to dissemination.

There may lie the answer as to why the engine and boilers remained where the ship grounded near a Halifax wharf. The conflicting force of the blast from the fore and after holds could perhaps have served to neutralise the eviction-effect and left these objects in place. Of course the sheer size and weight of the machinery may have been beyond the capacity of the explosive force to displace.

Many large pieces of steel landed on the Dartmouth side of the Harbour and those that fell on roads were the ones most visible. Charles Duggan a harbour ferryman with his own passenger boat was a reliable observer and reporter of the explosion and its affects. He had tried to approach ***Mont-Blanc*** with his ferry after the collision. In his escape by foot along Windmill Road on the Dartmouth side of the Harbour he described: "Twisted bits of the Mont-Blanc, scattered over the streets, made the run treacherous." Others described the shrapnel from ***Mont-Blanc*** as ". . . a hailstorm of metal slashing through the air." One 600 lb (75kg) piece of ***Mont-Blanc*** metal tore open the fence surrounding one of the military magazines.

The eye and part of the arm of a swing arm lifeboat davit showed a great deal of surface damage suggesting that the broken-off steel violently encountered many similar jettisoned objects during its trajectory from the vessel.

Virtually all the scrap-metal from the ship was cleared away in the following days and weeks and eventually made their way to dumps or even perhaps steel mills. Only a relatively few items were handed in to the museum but these have shown by their variety and condition what can happen when an enclosed cataclysmic explosive force acts.

Some final notes of interest are the different ways in which some of the artefacts were presented to the museum, and the variety of information about the *finds*. One had a small embossed plaque pop-riveted onto the plate –with wrong date and misspelled ship's name. Others were drilled and screwed onto wooden plaques. The places where they were found differed from a simple "dug up in the garden," to embedded in the rafter found when the roof was being repaired. Perhaps strangest was a small group of metal pieces which were collected and welded into a sculpture which was intended as a memorial,

Each annual anniversary is remembered in two significant ways: A Christmas tree is sent to the city of Boston, MA, USA in memory of the immediate help that city sent to the beleaguered city of Halifax. A train load of medical supplies and the staff to use them, doctors and nurses was sent to Halifax to help the recovery within days of the explosion. And a commemorative service is held at the memorial belfry erected overlooking the site. The bells came from destroyed city churches.

AUTHOR'S BIOGRAPHY

A retired Naval Architect now morphed into a Research Associate at the Maritime Museum in Nova Scotia. Specializes in historic ship preservation and conservation and the ethnography of historic provincial small craft. Instituted and conducted the Nova Scotia Small Craft Survey studying and recording the traditional provincial watercraft. Wrote the text to a photographic book, *Small wooden boats of the Atlantic (Provinces).*

Figure 8: PS: What is this?

STEM FOR ALL ON THE SS FRESHSPRING: VR TOUR

H. K. Yaqub, L. Star, BMT, UK
M. Kemp, J. Puddy, SS Freshspring Trust, UK

SUMMARY

Over the past year the world has changed dramatically. With greater restrictions on accessibility, the need to provide innovative and distributable remote experiences is now more prominent than ever. BMT has partnered with the SS Freshspring Trust to create multi-generational STEM experiences. "Preserving the past to inspire knowledge for the future". The SS Freshspring Trust have a vision to become a STEM hub by utilising cutting-edge technology. BMT have extensive experience in developing VR applications in the Maritime Domain. With skills shortfalls in many engineering disciplines, there is a need to inspire future generations into careers in STEM. Equally, many adults have a passion for technology and have valuable skills to offer to STEM projects. This paper uses the historic vessel SS Freshspring, a 1940s RFA Fresh Water Carrier currently being restored in North Devon, as the basis for exploring a range of initiatives and activities aimed at making engineering and technology interesting and accessible to all. The specific focus is on the development of an interactive 3D virtual tour, aiming to provide access to a wide audience by targeting a range of modalities including smartphones, internet browsers, and most consumer VR headsets.

NOMENCLATURE

AR	Augmented Reality
HMD	Head-Mounted Display
VR	Virtual Reality

1. INTRODUCTION

With the current COVID-19 countermeasures leading to greater accessibility restrictions, the need to provide innovative and distributable remote experiences is now more prominent than ever. The delivery of Science, Technology, Engineering and Mathematics (STEM) and heritage educational experiences often take place in the classroom, onboard the ship or at organised events. With all options facing new rules and restrictions, the traditional delivery of STEM is at risk and revised routes for engagement must be considered.

The SS Freshspring Trust provide educational tours onboard their steamship based in Bideford, Devon. The intention is for the heritage vessel to become a hub of novel and engaging STEM activities. The necessity for facilitating remote solutions for STEM and heritage education has led the Trust to rethink their engagement offering and how it can be delivered.

Since the Second World War, the world has not faced such a profound need to achieve and deliver innovative development. During this challenging time, allied ships were being sunk faster than the rate they could be produced. As a response to this, the industrialist Henry Kaiser introduced new techniques such as modularisation and welding which significantly shortened the production time, therefore revolutionising the shipbuilding industry. [1] These new methods became the foundations for modern shipbuilding which are still in use today.

Throughout the pandemic, the media has shared stories of innovative engineering which has given an insight to how STEM can be utilised to solve real-life problems and make a difference. The unprecedented circumstances have led to new partnerships being formed [2], significantly shortened design periods [3] and engineers openly sharing designs for manufacturing inexpensive medical equipment [4]. This current situation, although troubling, provides a unique opportunity for UK engineering to become a world leader. For example, a very successful collaborative project by University College London, Mercedes and Ventura has created a continuous positive airway pressure (CPAP) breathing aid. This has been delivered to hospitals all over the country with the first prototype deployable within only 100 hours [5].

As a purposeful and determined response to the pandemic, the SS Freshspring Trust have partnered with BMT and the University of the West of England (UWE) to translate the traditional activities into digital solutions which can be delivered remotely. The team are working to create a VR application which will not only allow 'remote tourists' to take a guided tour of the ship, but also participate in virtual activities. To ensure the Trust's inclusive ethos is maintained, the application is being developed to target a range of economical, lightweight and portable modalities to make it more accessible to all.

The long-term aim of the project is to provide a sandbox environment consisting of an interactive guided tour of the SS Freshspring which uses novel media presentation. These include interactive activities, immersive viewing experiences and freedom of exploration of the ship including areas which are not easy for visitors to access otherwise. The team's goal is to provide an experience in accordance with their mission statement: 'STEM for all' providing content to engage and inspire users of all abilities and ages.

2 THE TEAM

2.1 Project Formation

The SS Freshspring Trust was formed in 2013 to save the steamship from being scrapped. The Trust fostered a collaborative relationship with UWE in 2015 to provide UWE students with the opportunity to carry out design work on the conversion of the ship to carry passengers. In 2016 the Trust gained a £155,000 grant from the National Heritage Memorial Fund which enabled the ship to be docked for repair and moved to her home base of Bideford. As the partnership with UWE developed, it led to the creation of a STEM primary school project. The STEM project has been successfully operating in North Devon Schools for five years and has led to the development of a children's website. This work is ongoing within the partnership with a rollout planned in the new year.

The Trust has now set up a new partnership with UWE and BMT for the creation of a suite of VR activities based on the SS Freshspring. This has been driven by the desire to continue delivery of activities and tours in a safe way that ensures positive ongoing engagement with heritage. As high-quality consumer VR becomes more affordable and widely available, the innovative project will provide high value educational experiences to many and inspire young people to become engineers of the future.

The idea for a virtual tour originated from the Trust identifying the need to provide a means for visitors with limited mobility to participate in a tour of the SS Freshspring. The partnership team agreed that an immersive VR-based application could be achieved by capturing high quality 360-degree footage onboard the SS Freshspring using specialist equipment on a well-managed limited budget. The project complements BMT's ambition to offer "STEM at Home" activities with the aim to provide accessible extra-curricular challenges for students to participate in remotely. BMT's capability for developing distributable activities combined with the Trust's experience in building novel multi-modal engineering activities provides the perfect opportunity to curate a virtual guided tour which can be experienced in an engaging and immersive way.

2.2 Team Experience

2.2(a) BMT

BMT provides independent technical expertise and consultancy, offering design solutions, asset management, programme delivery and technology services to customers in the defence, security, transport, energy and infrastructure markets. BMT run a STEM outreach program to engage and inspire future talent and have extensive experience in developing engaging and novel activities. BMT's STEM offering features hands-on activities such as building foam boats, a 'Stability Ship' to demonstrate the impact of cargo loading on stability and a modular 3D-printed engine room which demonstrates the engineering principles of a ship's propulsion system.

BMT have developed a collection of VR and AR applications which allow the user to be immersed in novel environments such as space and at sea. In response to the pandemic, BMT have been looking at ways of delivering existing STEM events and sessions remotely including developing a suite of AR-based "STEM at Home" activities such as a mobile-based version of the propulsion system activity as seen in Figure 1.

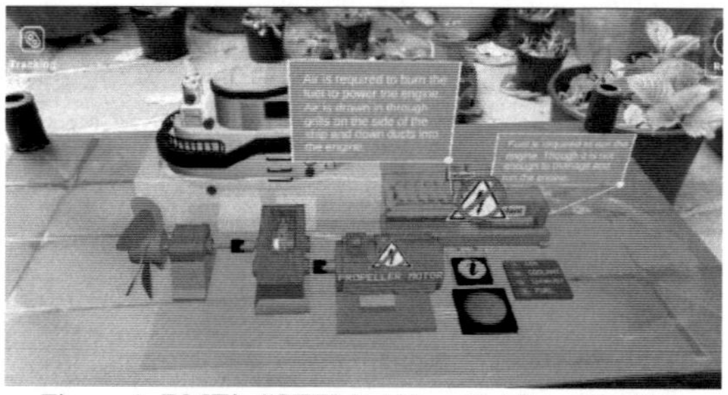

Figure 1: BMT's "STEM at Home" AR application. Users are required to correctly connect the engine components to the corresponding ship systems.

2.2(b) SS Freshspring

The Steamship Freshspring Trust is a not-for-profit organisation based in North Devon, whose aim is to return SS Freshspring to full operation, preserving the past and inspiring knowledge for the future. [6] The Trust has considerable maritime experience, both in renovation of a heritage vessel and managing projects.

The Trust offers a STEM Engineering Box Project, seen in Figure 2, [7] for local schools which is targeted at key stage two. The activity consists of six linked lesson plans which are delivered in sequence to demonstrate basic engineering principles. The lessons allow students to investigate and create a range of mechanisms which are used in most engines. On completion, pupils can visit the steamship and demonstrate their learning by identifying the same mechanisms on the steam engine. Other activities include ropemaking, creation of an interactive children's website, educational ship tours, flag lessons and an introduction to technical drawings.

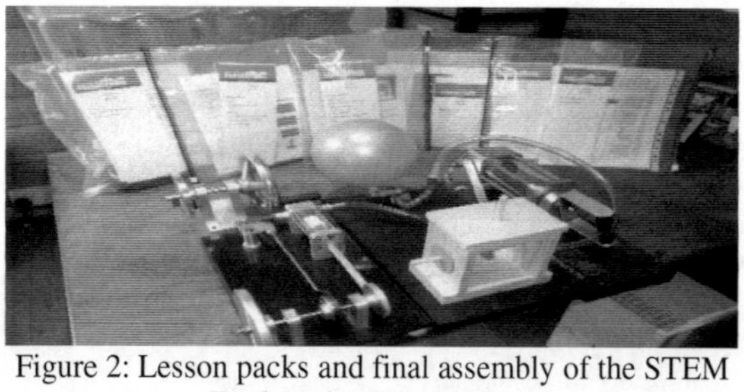

Figure 2: Lesson packs and final assembly of the STEM Engineering Box Project.

3. THE VIRTUAL TOUR

3.1 Footage and Content

The virtual tour enables remote visitors the freedom to explore the SS Freshspring in detail by allowing them to navigate between various viewpoints. The current tour has been constructed using 360-degree images and videos captured throughout the vessel. Interactive elements (hotspots) are overlayed points of interest which allow users to view additional information or navigate to another viewpoint. The camera used (Insta 360 One X [8]) facilitates rapid capture and processing of 360-degree footage, therefore enabling fast iteration of an initial prototype and minimising up-front production costs.

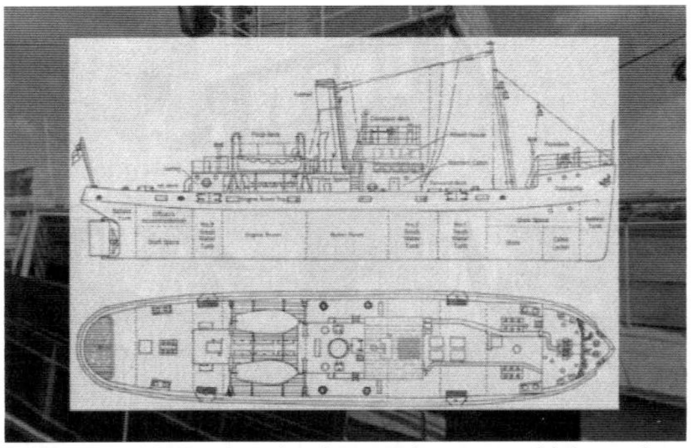

Figure 3: SS Freshspring interactive map which can be used to navigate to different viewpoints

Utilising 360-degree footage for the tour, allows an effective prototype to be made whilst keeping the hardware requirements to a minimum. This approach ensures the tour is developed in accordance with the mission statement 'STEM for all'.

The viewpoints for the 360-degree footage capture were pre-planned to ensure the user is given sufficient manoeuvrability within each area of the ship. The general arrangement of the SS Freshspring can be seen in Figure 3. Additional 360-degree video clips were captured depicting actors undertaking various activities which will feature in the tour (see Figure 6). The virtual tour grants the user access to ship features which are often out of bounds to visitors such as the boiler room, the water tanks and the propellors.

3.2 Functionality

The method in which the tour is implemented allows for multiple methods of interaction which include mouse-and-keyboard, touch, gaze-based interaction and VR controllers. This ensures the same package can be accessed through multiple means, allowing participation from a wider audience. Locomotion through the ship is undertaken in the style of "Google Street View" which uses the same principle of transitioning between two 360-degree images. Users can use an interactive map to navigate to different viewpoints as soon in Figure 3.

The targeted platform is the Oculus Quest 2 [9], which is a standalone VR headset which can operate without being connected to a computer. VR headsets like the Quest 2 allow users freedom of movement within an arbitrary physical space and offer comparable performance to its PC-based counterparts at a significantly lower cost. With greater affordability it easy to provide an immersive educational tour experience to institutions such as schools, retirement homes and community centres.

The application is being implemented for a number of modalities which include VR, web and mobile platforms. The method of interaction will differ for each modality (i.e. VR controller, mouse-click, screen tap and timed gaze selection). The application has been designed to only require basic one-click interactions, ensuring ease of deployment to different platforms.

Supplementary material is supplied in the form of ambient audio clips and hotspot within the tour denoting points of interest which allow the user to examine it in greater detail (see Figure 5). For instance, when they navigate to the wheelhouse, they are able to interact with points of interest placed in the scene (see Figure 4). Another feature of the tour includes the ability to visualise and compare SS Freshspring's three-cylinder triple expansion engine with a modern internal combustion engine to identify the key differences between the two. Other comparisons are currently being implemented.

Figure 4: Screenshot of VR Tour showing interactive hotspot and navigation button (Wheelhouse)

Figure 5: Screenshot of VR Tour showing interactive hotspot and navigation button (Lower Engine Room)

3.3 Narrative

While users are able to freely explore the ship, a guided tour is being developed to create a carefully curated experience suitable for all ages and abilities. During the virtual tour users will visit many of the ship's compartments, learn about Freshspring's history (and others in its class) and experience something of what living and working onboard the ship would have been like for the crew.

The narrative will take the viewer on a journey from the quayside in Bideford, back in time to the late 1940s to join Freshspring on a voyage to the Mediterranean where she has been tasked to resupply Royal Navy ships with purified water in Malta. Events during the voyage are drawn from real-life records, not only of the Freshspring, but also from her sister ships in the Fresh class which operated throughout the war. This will allow the story to expand and include the war years as well.

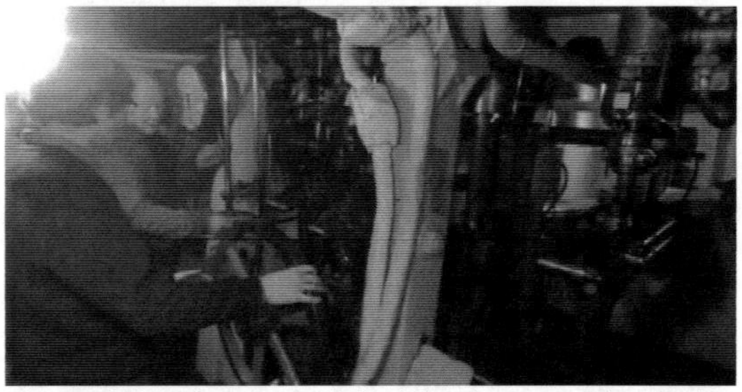

Figure 6: Supplementary 360 video demonstrating engine operation

The experience starts on deck as the ship is about to cast off. During the voyage, the viewer can visit different areas of the ship and meet members of the ship's crew, played by actors, who will carry the story forward. During the developing story, the perspective of the user can break out into different film or VR assets as seen in Figure 6. A scenario may consist of an engineer in the engine room reminiscing on how life on board was very different when the ship was coal fired. There would then be a following sequence with stokers shovelling coal filmed onboard a working steam ship. Events on the voyage add further flavour to the virtual journey until finally the ship arrives in Malta and the Freshspring begins her vital work as one of the Royal Navy's fresh-water carriers.

4. FUTURE ENDEAVOURS

4.1 Fully Realised VR

As more funding is secured, the project will expand to include more 3D interactions such as fully navigable 3D versions of compartments such as the bridge and engine room. This will also allow for the development of immersive activities which can be associated with different parts of the ship where users can participate in hands-on learning tasks. The benefit of a full 3D reconstruction of the vessel is that it will provide a more immersive experience and make navigation around the vessel feel more organic as opposed to the "Google Street View" solution currently being used for navigating between static viewpoints. The tour will also benefit from the inclusion of a realistic physics and animation system which will offer a naturalistic feel to the tour and activities.

This tour will act as the basis for a more activity-laden sandbox which is tailored to different skillsets, education levels and interests. With fully 3D assets, more interactive elements can be implemented including an engine which users can manipulate and operate. Additional challenges can be added during the simulation, for instance, a flood in the engine room or heavy seas that require a change of speed and direction. Additional assets can be introduced which may help the crew to decide how best to solve a problem, for example, a short video featuring a chief engineer who could retell a real-world experience providing guidance to the activity.

4.2 A Mixed-Reality Steamship

The development of assets for a fully 3D walkthrough can also be utilised for more ambitious mixed reality projects. Mixed reality refers to the use of AR and virtual assets in a real-world setting which uses tangible interfaces. For example, steam and water flow can be visualised directly through the pipes and machinery in the engine room. This visualisation changes according to which valves have been opened or closed. When members of the public are able to visit the SS Freshspring in the future, it will be important to provide new and exciting digital activities which fully utilise the ship itself. The virtual tour will still play a very important part catering for members of the public with limited mobility who may not be able to explore all parts of the ship.

4.3 STEM Outreach Partnerships

Once the virtual tour is complete and available to use, the project team aspire to facilitate remote STEM engagement via multi-generational routes to ensure the 'STEM for All' mission statement is achieved. By promoting via various social media channels and partnership websites, engaging with groups such as universities, schools, organisations and special interest groups, the team hope to actively engage with all generations and abilities by spreading awareness and creating new partnerships.

5. CONCLUSIONS

When the project began, the virtual tour was designed for users with limited mobility in mind. Since COVID-19, the project has evolved into a tour to provide users all over the world with remote access to the SS Freshspring. The unexpected circumstances of the pandemic have provided a unique opportunity to showcase the SS Freshspring globally as an asset to maritime heritage and its significant potential for STEM outreach.

'STEM for all' is a major aim of this project. People of all ages will have the opportunity to engage and learn about innovative technology; how it relates to heritage, how it can be utilised to create learning tools, how it is part of the future and how easy it is to access and use. Using platforms and tools that combine new and old technology for STEM delivery will also encourage different generations to experience the environments and activities together. Younger generations can learn about the principles and foundations of modern technology which itself can also be demonstrated in a simple way to older generations. STEM should be exciting and challenging for people of all ages and abilities. The potential of a distributed activity which can be viewed on a VR headset, browser or phone, will allow a greater potential for the general public to engage positively and benefit from STEM subjects. The project team hopes this will inspire and encourage users to participate in more challenging projects and provide them with a greater understanding of engineering and improved confidence in their own ability to undertake STEM tasks.

The aim of this project is to create an immersive and interactive guided tour of the SS Freshspring and its journey through history whilst providing an engaging sandbox of STEM activities for all ages and abilities. The project provides great potential to inspire STEM learning and facilitates the opportunity for users to consider possible career opportunities in engineering.

When pandemic-based restrictions to the SS Freshspring are lifted, the Trust hope to run onboard tours again. The VR project will provide a foundation for recreating the tour to include mixed reality activities and experiences onboard the vessel itself.

6. ACKNOWLEDGEMENTS

BMT STEM Ambassadors for their enthusiasm and commitment to providing young people with engaging opportunities to inspire their future choices.
Heritage Emergency Fund for facilitating the purchase of the specialist equipment required to make the virtual tour possible.
National Lottery Heritage Fund for supporting the Trust in both their educational work and conservation of the vessel.
Nautilus International for their support to this project.
SS Freshspring Trust Volunteers for their on-going and invaluable support. Without their dedication and hard work, the ship would not have survived.
UWE for fostering a partnership which has led to the growth of both the Trust's STEM and conservation work.
Worshipful Company of Shipwrights for their support to this project.

7. REFERENCES

1. THE ANTIPLANNER, Henry J.Kaiser: The War Years, 2008.
2. BEALES, A., How can Partnerships Create Rapid Innovation during COVID-19 Crisis? 2020
3. JEBERA, P., Dyson Designs Ventilator in 10 days for COVID-19 Patients. 2020
4. NORDRUM,A & STRICKLAND E., Engineering during the Coronavirus 2020
5. ULC INSTITUTE OF HEALTHCARE ENGINEERING, UCL Ventura breathing aids for COVID-19 patients, 2020
6. SS FRESHSPRING TRUST, Steamship Freshspring Trust. 2020
7. SS FRESHSPRING TRUST, Engineering Box Project Overview, 2020.
8. INSTA360. Insta360™ ONE X Own the moment, 2020.
9. OCULUS QUEST 2, Our Most Advanced All-In-One VR Headset, 2020.

8. AUTHORS BIOGRAPHY

Dr Hashim Yaqub holds the current position of Visualisation Developer at BMT. He specialises in developing VR and AR applications as well as working various research projects ranging from their application in training to engineering design. Hashim has 7 years of experience in his field and is responsible as a technical lead for the VR tour.

Martin Kemp holds the current position of Creative Media Director for the SS Freshspring Trust. He is responsible for designing, producing and directing the Steamship Freshspring Trust's 360 video and Virtual Reality projects. His previous experience includes more than 30 years' experience as a factual documentary director for UK and US television channels including BBC, Channel 4, National Geographic and Discovery. Specialising in history, science and current affairs on film locations around the world. Previous projects include 3 months filming onboard the Royal Navy's icebreaker, HMS Endurance, in Antarctica. Extensive experience in using CGI for documentary. Martin's academic background is in physical sciences.

John Puddy holds the current position of Trust Chair of The Steamship Freshspring Trust. He is responsible for Managing the Governance of the Trust and overseeing projects. His previous experience includes Ship Management and operations, past Director of Amber Foundation, working with young, disadvantaged groups, past Director of Greenpower a STEM related school project and Founder of the Steamship Freshspring Trust.

Laura Star holds the current position of Graduate Consultant within Naval Architecture at BMT and is a BMT STEM lead. She sits within the Outfit and Arrangement team and is responsible for supporting design solutions to specialist ship design challenges. Her previous experience includes weights management for Ship and Submarine designs.

THE *ALBUM DE COLBERT*: THE IMAGE OF SHIPBUILDING

M. Corradi, Università degli Studi di Genova, Italy

SUMMARY

The *Album de Colbert* compiled by an anonymous author in the second half of the seventeenth century is among the most important illustrated testimonies of the art of shipbuilding. Probably commissioned by Jean-Baptiste Colbert, Minister of Finance of the Kingdom of France, it was compiled with the aim of making Louis XIV understand the complexity of shipbuilding. The *Album* was made to support the creation of a navy with the ambition of being competitive with the Royal Navy and with the intent of modernizing and expanding the French shipbuilding industry. The fifty plates that make up this illustrated treatise unravel the story of the construction of a first-rank 80-gun line vessel, from the laying of the keel to the launch. It is a unique document that has no contemporaries or precursors because it is not a didactic collection of boats, like the previous treaties that had a completely different methodological approach, more technical-descriptive than illustrative, but wants to go beyond the scientific treatise. Its purpose is instead to measure itself with representation, showing through the strength of drawing and images the peculiar aspects of the reality of shipbuilding, using iconography as a means of transmitting knowledge related to the world of shipyards and shipbuilding in the 17th century.

1. INTRODUCTION

The *Album de Colbert* [1], now preserved at the Historical Navy Service in Vincennes, compiled by an anonymous author in the second half of the seventeenth century, is among the most prestigious illustrated examples of the art of shipbuilding [2]. It was drafted probably on behalf of Jean-Baptiste Colbert (1619 - 1683) [3], Minister of Finance of the Kingdom of France, at the time of Louis XIV of Bourbon, known as *Le Roi Soleil* (1638 - 1715). It was composed to illustrate the construction of vessels in France, and, indirectly, to testify to the desire to create a navy with the ambition of being competitive with the Royal Navy, if not even becoming the first navy in the world. This project was a consequence of the economic and colonial policy of the French sovereign, a policy suggested by his Minister Colbert: having control of the sea, of the commercial shipping lines and of military dominance, implied having a permanent war fleet, to guarantee the French maritime interests, as it had already been imagined at the time by Cardinal Richelieu (1585 - 1642). In 1624, Richelieu decided to equip the kingdom of France with a real state navy and created the first three French naval teams with bases respectively in Brouage, Brest, and Le Havre. Richelieu, designated as «grand maître, chef et surintendant de la navigation et commerce de France», also decided to make Toulon the main French military base in the south of the country, also because beyond the Strait of Gibraltar, ships under the English flag, Dutch and Hanseatic had a strong dominion over the seas. Navigation in the Mediterranean was then still a compromise between sailing propulsion and the muscular strength of the rowers made by the slaves embarked in the galleys. The latter moved from Marseille to Toulon to more easily counter the Barbary corsairs settled in the Hyères islands, while, beyond the Pillars of Hercules, the main ship was the vessel. Richelieu had already established that warships were no longer owned by local captains, ship-owners, or lords, but instead were owned by the State that had built, armed, and kept them in service.

In 1637, the French fleet included about forty ships, but the shortcomings of the royal navy were numerous: the officer corps was not very disciplined and had to be renewed with each military campaign. The recruitment of the crews was very summary: the sailors were men gathered in the ports and embarked by force, as was also the case in England. In 1640, Cardinal Jules Raymond Mazarin (1602 - 1661) was forced to considerably reduce the navy budget and from that moment the French monarchy's commitment in the naval field was secondary, and budget priority went to military operations on land. In 1642, 65 ships and 25 galleys were moored in the port of Toulon, almost the entire French fleet in the Mediterranean. Five years later, in 1647, the first "modern" vessel, the *Reine*, was launched from the shipyards of Toulon. Nonetheless, in 1661, the royal navy was limited to about thirty poorly equipped ships with little maritime activity. The decisive change of policy, therefore, took place with Louis XIV, inspired by Colbert, who wished to equip the country with a naval force capable of opposing the United Provinces (the Dutch Navy) and the England, with the aim of being able to conquer the control of the seas with a decisive numerical superiority of warships. In 1667, the first 120-gun vessel, the *Royal Louis*, was launched in Toulon. Two years later the Arsenal of Marseilles was completed and Colbert returned the galleys to Marseilles while he left the vessels in Toulon. In 1675 the galleys increased from twelve to twenty-five units. Their growth is linked to the fact that they were in any case an undeniable element of prestige and a powerful factor of intimidation, as well as a formidable weapon of combat, especially against Barbary pirates.

In this sense, the *Album* is likely to have been meant for Louis XIV with the intention of illustrating to the King the complexity of shipbuilding and therefore the need to modernize and expand the French shipbuilding industry, as well as starting a fleet renewal with the construction of a more modern ship in keeping with the rise of the

French state. The fifty illustrations ("*planches*"), that make up this illustrated document, unravel the history of the construction of a first-rate line vessel with 80 or perhaps 84 guns, from the laying of the keel to the launch.

It is a unique document that has no contemporaries or precursors because it is not a didactic collection of boats, like the previous treaties, that had a completely different methodological approach, more markedly technical-descriptive. It does not find followers even in the most famous and in some ways exhaustive *Encyclopédie ou Dictionnaire raisonné des sciences, des arts et des métiers* by Denis Diderot (1713 -1784) and Jean-Baptiste Le Rond d'Alembert (1717 -1783). It was superior for quality of the illustrations to works more dedicated to naval art, such as the *Architectura navalis* (Ulm: Jonam Saurn, 1629) by Joseph Furttenbach (1591 - 1667), the *Hydrographie, contenant la théorie et la practique de toutes les parties de la navigation* by Georges Fournier (Paris: Michel Soly, 1643), the *Elemens de l'architecture navale ou traité pratique de la construction des vaisseaux* (Paris: Charles-Antoine Jombert, 1752) by Henri Louis Duhamel du Monceau (1700 - 1782) or *L'architecture navale: contenant la maniere de construire les Navires, Galeres & Chaloupes, & la Definition de plusieurs autres especes de Vaisseaux* (Paris: Jean de la Caille, 1677) drawn up by François Dassié (17th century), drawing master at the school of the Design of Toulon, and again the work of Carel Allard (1648 - 1709), Dutch cartographer and engraver, who published in 1697 *L'art de bâtir les vaisseaux* [4], which takes up the encyclopaedic text by Nicolaes Witsen (1641-1717) [5] and the most descriptive treatise on Cornelis van Yk (1649 - 1712) entitled *De Nederlandsche Scheeps-Bouw-Konst Open Gestelt* [6], to name just a few of the best known in literature.

The *Album de Colbert* aims to go beyond the scientific treatise and instead, it measures itself with representation, showing through the power of drawing and images the peculiar aspects of the reality of shipbuilding, thus it uses iconography as a means of transmitting knowledge related to the world the shipyard and the construction of the vessels in the 17th century. The *Album* is an important dissemination work, also oriented to the transmission of technical and construction knowledge in the field of shipbuilding. The work was drawn up thanks to the foresight of Colbert, architect of a renewal of the arsenals and ports (Dunkerque, Le Havre, Brest, Rochefort, Port-Louis), as well as the French Navy, and promoter of the birth of the Naval Engineering Schools of Rochefort, Dieppe and Saint-Malo. An avant-garde character for his time, a visionary with his feet firmly on the ground that France only met too late. With his son, Jean-Baptiste Colbert (1651 - 1690), Marquis of Seignelay, Minister Colbert invited numerous craftsmen, shipwrights, and master rope makers from the Hanseatic cities and the Netherlands to work in France to install and set up arsenals, and shipbuilding yards in the kingdom's main ports.

In just ten years, the number of units built was quadrupled, from 30 to 120 ships: the French arsenals were no longer able to support by themselves the number of orders for new vessels, and therefore it became necessary to buy ships from Holland. This progression by leaps and bounds was accompanied by the training of naval personnel: a corps of naval officers was created, a system of recruiting crews was established, crews became more and more numerous. The training of naval officers initially took place in the royal schools of hydrography and was identical for both merchant navy and navy officers. But while merchant navy officers were chosen from seafarers belonging to the private merchant navy, the officers of the navy came from the nobility, because it was thought that their nobility would be enough to guarantee their abilities. Colbert chose Dieppe and Le Havre to establish the first hydrography schools.

A royal ordinance of 1681 defined the subjects of teaching, recruitment, and the duties assigned to officers and non-commissioned officers on board ships. Ordinary citizens went to hydrography school to get a job in the merchant navy but if necessary they could become "auxiliary officers" in wartime. A young man was introduced into this training activity from the age of twelve as a "novice". Then, when he was fifteen years old, he became a "volunteer" and he began his career as a naval officer. Finally, Colbert favoured the emergence of important commercial companies such as the *Compagnie des Indes Orientales*, the *Compagnie des Indes Occidentales*, and the *Compagnie du Levant*, along with an expansion and renewal of the merchant navy. After the siege of La Rochelle (1627-1628), the end of the Thirty Years War (1618-1648), and of the Fronde (1648-1653), the situation of the French merchant navy was catastrophic.

The naval force that Richelieu had assembled - 80 ships and 20 galleys - had been almost destroyed during the Fronde. Arriving at the Ministry, Colbert found only 20 intact warships, of which only two or three could be used in navigation on the high seas; out of 20 galleys, only six were still in service. In the event of naval conflict, therefore, France was obliged to charter warships abroad. In 1669 Colbert persuaded Louis XIV, who was barely interested in the maritime problem and rarely visited shipyards, to address the issue of navy renewal. Colbert obtained the authorization to create the Ministry of the Navy, of which he was the first Minister and which made him the father of the modern French Navy, a position that allowed him to take over the problem of the navy. In this capacity as Minister Colbert embarked on an ambitious modernization program for the French fleet [7]. To initiate this important Navy renovation project and interest the King, Colbert ordered to build a flotilla of smaller-scale ships, which he placed in the large body of water overlooking the palace of Versailles, and where the king could constantly admire the evolution of these ships. Subsequently, having obtained the interest and the approval of the King, Colbert started a shipbuilding program, freeing France from the constraint of import ships from abroad.

2. COLBERT AND THE BIRTH OF LOUIS XIV'S NAVY

Colbert shared the opinion of the engineer Louis Nicolas de Clerville (1610 - 1677), who considered very harmful the habit of «chartering Dutch ships rather than building them ourselves» [Letter from Clerville to Colbert of 18 May 1663]. At the time of Colbert's entry at the Ministry of the Navy, France had a rather modest fleet. The investigation ordered by Colbert and conducted on 17 May 1664 in La Rochelle by the Admiralty judge Jean de Mirande (1630 - Unk) showed the inadequacy of the merchant navy stationed in La Rochelle. Of the 32 ships in the port, only 18 had a tonnage of over 100 tons and only two of these were between 300 and 400 tons. In the rest of the country, the situation was no better; there were, in fact, in all of France only 19 ships with a tonnage between 300 and 400 tons, and then many other ships but of modest tonnage. The largest ships barely reached 400 tons. In addition, the ships were located in numerous ports: La Rochelle, Bordeaux, Nantes, Dieppe, Dunkerque, Abbeville, Honfleur, Calais, Le Havre, Saint-Malo, Rouen, Saint-Brieuc, Bayonne, Marseille, Toulon, and Saint-Jean-de-Luz. They were basically small boats (flûtes), flat-bottomed, large and heavy, with a round stern, the most used merchant ships at that time. All these ships were weakly armed and had no other use than for the transport of goods. They were also relatively old ships, most of them over sixteen years in service and some as long as twenty. Few were the ships under ten years of activity, and also the majority had been built abroad. In addition, the French Navy still had many "d'antan" ships, galleys concentrated in Marseille and Toulon to form the naval squadron of the Mediterranean.

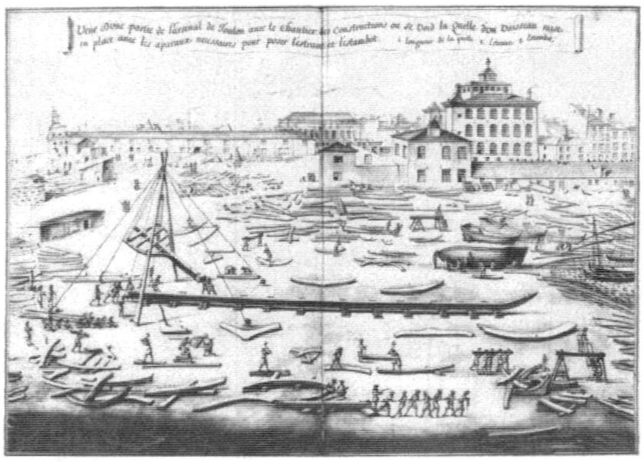

Figure 1: View of the Arsenal of Toulon, from *Album de Colbert*, Plate n. 1.

In a note presented to the Conseil du Commerce, Colbert pointed out the tragic situation of the French fleet and ports, in particular La Rochelle, which had been severely hit by a dramatic siege during the Thirty Years War, and he remarked to the King that the fleet was in conditions of great inferiority compared to the Dutch and English ones. On August 10, 1678, France signed the peace treaty of Nijmegen, at the end of the war against the United Provinces. By winning, France took control of Franche-Comté, some Flemish cities, and the county of Hainaut, and Louis XIV was nicknamed Louis the Great. It was an important victory in a war, the so-called Dutch War (1672-1678) fought by the Kingdom of France against a coalition formed by the principality of Brandenburg, the Holy Roman Empire, Spain, and the United Provinces. French Navy also fought on the sea, for the first time in the history of the kingdom of France, and it achieved decisive successes during the conflict, and the nickname acquired by Louis XIV is due to the merits achieved in battle by Marshal Henri de La Tour d'Auvergne, Vicomte de Turenne (1611 - 1675), by Admiral Abraham Duquesne (1610 - 1688), winner in the Mediterranean of Dutch Admiral Michel Adriaenszoon de Ruyter (1607 - 1676).

The successes in the naval field meant that Colbert commissioned his famous iconographic album, which was to represent the construction of a vessel, an instrument of power and hegemony on the seas, but also a symbol of the military power of the French State. His aim was to show the king the complexity of building warships, in order to arm the fleet of France and make it equal to that of the English, Spanish and Dutch powers [8]; establish a state of art knowledge in shipbuilding in order to initiate a navy renewal and expansion program. Furthermore, his goal was also to educate his son Jean-Baptiste Antoine Colbert, Marquis de Seignelay (1651 - 1690), and soon his successor, in the burdensome duties as Secretary of State for the Navy (1683), and later Minister of France. (1689) [9].

Colbert in a short time became increasingly aware that it was necessary to provide for the construction of a fleet "on his own", as he could no longer depend on foreign shipbuilding. For this reason, he promoted the renewal, modernization and expansion of shipyards that allowed him to create a shipbuilding industry capable of producing ships in series on the basis of a standardized project. However, in this standardization process each site had to be autonomous in terms of planning, management, site management, procurement of materials, etc. The Album de Colbert, which describes in detail each phase of the construction of a vessel, was, therefore, the synthesis of this ambitious program of the French Navy Minister. The principles of construction and armament of warships had to be such that naval squadrons could thus be competitive with those of the English and Dutch navies. This massive industrial effort, truly formidable for the time, obviously had to be implemented with substantial funding from the Crown. «Vous pouvez nous détruire des navires, nous avons amplement de quoi reconstruire» wrote Colbert about this. In 1662 Colbert obtained a sum of three million lire for the navy, a sum that the Minister was able to quickly double and then triple in a few years.

In 1670, funding of 13 million lire was even awarded, with an average annual indemnity of about ten million (five times greater than the two million lire spent by Mazarin shortly before his death). The results were not long in coming thanks to the impressive work carried out in the yards of Toulon, Rochefort, Brest, Le Havre, Dunkerque, Boulogne, and Calais: in 1671, the French Navy had 196 ships as well as a certain number of galleys; in 1677, it had 300 ships and galleys, with approximately 50,000 sailors in service, which will soon increase to 160,000. It was certainly an exceptional undertaking because the yards were devoid of everything: as regards construction materials, there was a lack of wood, ropes, tar and products for caulking, bronze and iron for small parts and accessories, guns and armaments in general, etc. Colbert wrote in a memorandum of May 20, 1671: «Non seulement à l'égard du goudron, écrit-il le 20 mai 1671, mais même de toute autre marchandise propre à la marine, il faut généralement s'en fournir dans le royaume, plutôt que d'en prendre des étrangers, quand bien même ces dernières seraient à quelque chose meilleur marché; étant important, pour mettre nos manufactures en valeur, de s'en servir de préférence à celles du dehors et de convier par cet exemple les marchands à s'en servir de même».

From 1669, the effort put in place by the Ministry of the Navy meant that the shipyards was amply supplied with everything necessary for the needs of a modern shipyard. Likewise, the ports and shipyards of Toulon, for the Mediterranean area, and Brest for the Atlantic area, became strategic. Brest also became the port par excellence where the fleet resided, which had to counteract the British and Dutch naval hegemony. The renovation of the construction sites also went hand in hand with an urban revolution of the port cities, and in this regard Sébastien Le Prestre, marquis de Vauban (1633 - 1707), military engineer and builder of fortifications, was called to collaborate, who participated in the expansion and to the defence of the port of Toulon where thanks to its port plan it was able to moor up to one hundred ships. The shipyard was expanded and equipped with everything needed to put numerous ships into production, about six or seven at the same time that had to be completed every year. In addition, the warehouses were replenished for the procurement of materials, equipment and for the construction, armament, and outfitting of ships. There were thousands of cannon pieces, and minor armaments, anchors, accessories of all kinds, sails and ropes in quantity thanks to the construction of new factories in the service of the Navy. According to Colbert's intentions, the new port-shipyard of Rochefort was to emulate, for production and landing capacity, the port city of Saardam, an impressive port in which the Dutch built their ships. The site chosen in Rochefort – "la Chatellerie de Rochefort" - was a village of about 500 inhabitants surrounded by marshes, and within a few years, thanks to the impressive works promoted by the Ministry of the Navy, it turned into a port city with a large shipyard where about 20,000 workers and technicians worked [10], so much so that due to its size and the wealth of industrial factories it took the name of "Versailles de la Mer". The architect François Blondel (1618 - 1686) drew up the town planning and development plans of the city and Louis Nicolas de Clerville directed the construction work [11]. Construction began in 1663 and seven years later, the new city was defended by a system of fortifications, and the shipyards were equipped with foundries for the production of guns, and gunpowder stores, dry and repair docks, weapons depots and food, factories for the construction of the masts of ships, an imposing factory of ropes (the royal rope factory), and factories for the production of what is necessary for the construction and preparation of warships. In addition, the city housed officers and engineers who worked on the plans for the new ships. The shipyard's production capacity was substantial: in 1673, 26 large ships and 6 galleys were built, along with other smaller boats [12].

The centralization of naval production in the arsenals and the standardization of the construction of new vessels made it possible to build in series the different types of vessels as well as frigates, minor ships and even galleys, which still found their use in the French Navy. This new production system brought considerable advantages in the management of the shipyard and in shipbuilding, with a significant reduction in construction times. The chronicles tell that in the port of Rochefort it was possible to set up a frigate in 30 hours, in Brest in 22 hours, while in Marseille a galley was set up in just 7 hours! During one of these performances, in Marseille, with Colbert's son present, the workers began to set up a galley starting at 6:30 in the morning, and the Marquis de Seignelay saw it sailing around 5:00 P.M. towards the Chateau d'If.

This enormous industrial effort was being accompanied by a great commitment in the training of shipbuilders, in their various specializations, as well as that of training newly recruited sailors. For this reason, Colbert made sure that a naval school was established in each port to meet the needs of ship building and armament. A maritime school and a hydrography school were also set up in Rochefort, as well as an artillery school; in Saint-Malo a Collège de marine to teach hydrography to naval officers and sailors, and in 1693, all the efforts made in particular in the field of hydrography led to the development of numerous engineers specialized in drawing up maps and atlases, a sort of geographical mapping of the coasts (the *Neptune français* [13]) the first of its kind, which covered a geographical area that went from Norway to Gibraltar. The maps will later be considerably improved thanks to the classification according to the reference system in geographical coordinates according to latitude and longitude [14].

3. SHIPWRIGHTS AND NAVAL ENGINEERS AT THE SERVICE OF *ROI SOLEIL*

To expand the fleet Colbert made use of numerous expert shipbuilders [15]: Rodolphe Gédéon (Unk - 1672) in the Toulon shipyard, Jean de Wert (17th century) in the Chantier de la Seudre, Laurent Hubac (c. 1607 - 1682) in Brest, and others [16], as well as numerous skilled craftsmen such as Flemish and English carpenters hired to increase the skilled workforce. Nonetheless, Colbert promoted a massive import of construction and finishing

materials, such as wood, pitch, tar, copper, anchors, from northern Europe; Pyrenean wood, Auvergne hemp, pitch, and resin from Gascogne.
From 1689 to 1693, France put into service an average of 17 ships per year, a unique performance in its history. By 1691, the arsenals operated at full capacity, and orders for the navy reached an unprecedented level. The Royal Navy had 154 vessels in service, including 102 first and second rate ships, while the naval team of galleys peaked at 40 in the Mediterranean, while another 15 were built at Rochefort. Due to their low artillery equipment, galleys were used for escort or reconnaissance missions, as their combat effectiveness was significantly lower than ships of the line.
Such an economic and productive commitment also affected the production of wood. The French forests still had age-old oaks, a legacy of the policy adopted in previous years for the construction of warships for the royal navy. In the mid-17th century the French Navy was reduced to a minimum, relying on a small number of warships and thus having the need, in case of conflict, to charter or buy ships from foreign powers. And in order to stop buying foreign ships in case of war Colbert decided to reorganize the entire production chain, from the cultivation of the oak, called Colbert's "treasure" [17], to the shipyard, and from that moment the kingdom's forests were preserved for military and economic interest. The construction of a large ship required, in fact, the felling of thousands of trees and this needed a prudent but above all planned forestry policy, in order not to excessively reducing the wooded areas with excessive exploitation. At that time, coppice was cultivated, consisting of young trees that grew back on the cut stumps, at least every 25-30 years, to supply the cities with wood and the salt and glass industry with fuel. Shipyards needed large pieces of wood, but not necessarily straight; in fact, the rounded shapes of the ships required curved trunks, sometimes with an unusual if not bizarre shape, due to particular construction elements. A first-rate vessel required nearly 3,000 oaks and a number of other species for its construction, such as fir or spruce for the musts, ash for the capstan, guaiac imported from the colonies for the blocks. The lack of construction timber had thus forced France to turn to foreign markets: oak wood was mainly imported from Italy and Albania; the pine from the north of Europe. Therefore, in the event of war, the control of the timber routes became strategic for a great nation like France.
In the eyes of the French statesman, however, the forest was an important source of wealth that had to be preserved and properly managed. The ordinance issued in August 1669 (Ordonnance de 1669 of Louis XIV «sur le fait des Eaux et Forêts» [18]) sanctioned a vigorous resumption of control of French forests. According to Colbert, it was necessary to preserve a quarter of the forest area; the age of the felling of the trees had to be reduced to twenty years, and for this reason, at least 32 young trees per hectare, among those judged by the forestry to be quite straight and vigorous, and which could become construction timber, had to be preserved in any case. In this way, precise rules were also established regarding the felling of trees. The results were not long in coming: the revenues of the royal forests were increased about twenty times in twenty years. Colbert also initiated the development of forests for the production of trees in Provence, Auvergne, and the Pyrenees. In this process of verification and renewal of the fleet, Colbert started the reorganization of the shipyards, and at the same time a process of standardization and industrialization that had to foresee the construction of models of vessels whose function was to serve as reference models in the shipyard, flanking the shipwrights of new professionals with good mathematical preparation, but above all of the geometry, to improve the performance of the new ships. The staff who directed the shipyard had to be a new professional figure who had to train in the naval schools; nevertheless, in the construction sites the employment of specialized foreign workers was not disdained, and if necessary they resorted to espionage to discover the secrets of the master builders of other countries.

Figure 2: How to use trees "naturally" to build a ship in the 18th century, from *Encyclopédie méthodique. Marine*. Paris: Panckoucke; Liége: Plomteux, 1783-87; planche 101 [21].

This project, which required a lot of skilled workers, both in the construction phase and in that of management and navigation, started, on the one hand, a new process of training for technicians and workers but also made sure that the best builders were invited to collaborate with the French shipyards. Europeans: the Neapolitan Biaggio Pangallo (Maître Blaise or Monsieur Blaise or Sieur Pangalo, 1650 - 1714 or 1719) [19] and the English Anthony Deane (1633 - 1721). The latter was accused of having transmitted military information to France; condemned and incarcerated in the Tower of London, he was finally released (9 July 1679 - 14 February 1680). Deane was one of the first to apply the scientific principles of statics to the construction of military ships between 1666 and 1675. Among others, the Dutch Heijndrick Dircksz Sluijck (1607 - 1686) also worked in France. Heijndrick Dircksz Sluijck was been one of the largest Dutch shipbuilders and timber merchants. He was overseer of the Waterlandse Doopsgezinde Gemeente (Waterlandse Baptist Church) and guardian of Zaandam.

In the years 1666-1667 six ships were built on behalf of Louis XIV: the *Neptun*, built by Jan Hendricksz Cardinael (17th century) in Zaandam, *Le Conquérant*, *L'Invencible*, *Le Courtisan*, *L'Intrépide* and *Le Normand*, the latter built by Hendrik Dircksz Sluyck in Sardam or Saerdam (now Zaandam). The last three vessels then changed their names to *Le Magnifique*, *Le Grand* and *Le Saint Louis* in 1671. The effort undertaken by the Ministry of the Navy was considerable. In 1673 alone, twenty-six ships and six galleys left the shipyards of Rochefort. In a very short period of time, the French navy had more than 300 ships and galleys in service (1677), and within a century it became a prestigious navy comparable to the English and Dutch [20].

4. THE *ALBUM DE COLBERT*

This *Album* [22], whose original copy is kept at the Service Historique de la Marine, represents in detail in fifty sheets the construction of a three-deck ship, from the laying of the keel to its complete rigging of sails. The album is anonymous and has no date (probably it was written around the years 1660-70); however, the representation of the Arsenal of Toulon in the first table allows us to establish that the text is prior to 1677 since appears the building to prepare tarred ropes, for the subsequent caulking of the ships, building which disappeared in that year due to a fire. The *Album* does not refer to any specific ship built in that period, which reinforces the hypothesis that it was written with a mainly didactic-illustrative objective intended to tell through images the different phases of the construction of an 80 or perhaps 84 guns vessel. In fact, between 1663 and 1667, eight ships of this type were built in the French shipyards, including the following four in the Arsenal of Toulon: *Saint Philippe* (1663, 74 then 84 guns), *Paris* then *Royale-Thérèse* (second rank vessel, 1670, 70 then 80 guns), *Île de France* then *Lys* (1691, 76 then 84 guns) and *Sceptre* (1691, 80 then 84 guns) [23].

The *Album de Colbert* is a unique work and differs in the quality of the images and completeness of the descriptions from contemporary works published in the 17th century in France. Among those in print, we recall the aforementioned *Hydrographie* by Fournier (1643), the Dassié's *Architecture navale* (1677), and *De la construction d'une gallaire et de son équipage* by Ithier Hobier of 1622 (Paris: Denis Langloys). The authors of the Album are anonymous. We are talking about authors because probably more designers and engravers have dedicated themselves to the editing of this volume, even if there are some similarities, in terms of style in drawing and shading, with a volume dedicated to the merchant navy, the *Albums de Jouve, Desseins de tous les bâtiments qui naviguent sur la Méditerranée,* par Jean Jouve, de Marseille published in 1679 and composed of 32 plates.

The Album is made up of 50 plates and is intended to illustrate the construction phases of a vessel. The ship represented in the plates is a first rate three-deck line vessel without a forecastle, armed with 80 guns, and probably with a tonnage of around 1.500 tons. The vessel represented complies with the Navy regulations, which defined the arrangement of artillery on three decks for those of the First Rate. The Navy Ordinance of April 1689 [24] is the one that set the dimensions of this category of ships for the first time: «... cent soixante-trois pieds de longueur de l'estrave à l'estabot par dehors; quarante-quatre pieds de largeur en dehors les membres, & vingt pieds quatre pouces de creux, à prendre sur la quille au-dessus des bouts du bau en droite ligne» [p. 292]. It was an imposing ship for the time: the length from the stern to the bow is 148 French feet (about 48.07 m / comparison with the Paris foot equal to 32.484 cm), maximum beam is equal to 40 feet (c. 12.99 m), and the draft is equal to 20 feet (c. 6.50 m).

A complex vessel from the morphological-structural point of view that required the best woods for its construction and built according to a model-type system called *maître-gabarit et de la tablette* [25]; this method made it possible to create the skeleton structure from the bow to the stern of the vessel by tracing construction lines (the water lines) starting from the main section that defined all the dimensions of the vessel, with a displacement suitable for the loads it had to support [26].

The vessel had the number of guns divided as follows for each side: the first battery deck had 14 gun-ports, as well as the second battery deck, while on the third deck there were 13 gun-ports (four forward and six aft) and 3 on the forecastle.

Figure 3: Plate n. 42 showing the vessel with its masts, yards, and rigging ropes.

If we refer to the rules issued from 1669 to 1674, the artillery of the vessels of the time should have been composed as follows:

- 1st battery: 12 x 24 lb cast-iron guns and 14 x 18 lb iron guns.
- 2nd battery: 14 18-pound iron guns and 12 12-pound iron guns.
- 3rd battery: 26 8-pound cast-iron guns.
- Stern Castle: 6 x 4-pound cast-iron guns.

The vessel does not refer, as mentioned, to a specific vessel, but refers to the ships then built at Toulon, however very similar to each other even if made by different shipwrights and of which we have mentioned: the *Saint Philippe*, a project by Rodolphe Gédéon, the

Paris then the *Royale-Thérèse* based on a project by Jean Serrin (17th century), the *Île de France* then the *Lys* based on a project by Louis Audibert (17th century) and the *Sceptre* based on a project by Laurent Coulomb (1622 - 1696), where there is a similarity in the decorations at the stern and even in the figureheads. Although the whole of the plates in the *Album* is expressive of the geometry of the 80-gun vessel, however, it should be noted that some plates are more artistic than technical.

The drawings are very beautiful and widely illustrated, but nevertheless, there are some inconsistencies or errors and the dimensions of the vessel as depicted are sometimes approximate. For example, if we try to overlap the plans of different decks, the openings and passages of the masts do not always coincide. Similarly, the profile (side view) in plate no. 39, the level of the stern and lower castle is different from that of plate no. 47, where the longitudinal section of the vessel is shown. The gun ports of the stern tower appear or disappear according to the different views. The list could be further extended, although it should be emphasized that it is possible that the Author of the *Album* observed the construction of several warships at different times, reporting in a single document the drawings of ships, however partially different from each other. The guns of the third battery are however quite weak and poorly protected. The first battery is very low compared to the waterline and therefore becomes unusable in the event of a strong heeling of the ship.

We will not go into the description of all the plates in order not to make the reading of this short note boring with a long list of data, even if, in fact, the *Album de Colbert* would deserve an in-depth analysis for the richness of the iconographic and descriptive contents. However, the length of this short note prevents us from doing so and we refer the curious reader to a more detailed examination of the *Album*. Plate no. 1 is a didactic image of the Toulon's shipyard, and it shows the keel set up with what is necessary for the implementation of the fore and aft stem. On the other hand, we like to point out how the ordered succession of plates from number 2 to number 38 represents the construction of the vessel step by step. Plate no. 39 shows the complete vessel at anchor. Plate n. 40 and plate n. 41 instead deals with the maintenance procedures of the hull, while plate no. 42 shows the vessel with its cage masts, flagpoles, and rigging ropes. The following plates (n. 43, 44, 45, 46, and 47) illustrate the interior fittings and the functions assigned to the individual decks, then plate number 48 showing a view of the stern of the vessel moored in the roadstead, plate no. 49 where you can observe the same vessel seen from the bow, and finally plate no. 50, the most iconographic of the album, which illustrates the vessel in all its magnificence.

5. CONCLUSIONS

At the end of this brief note, we believe that the *Album de Colbert* can be placed, with full rights, in that renewal of technical literature, which took place in the 17th century, and which will lead on the one hand to the didactic iconography of the great Encyclopaedias and from the other than the more technical ones that will give rise to specialized manuals. The work strongly desired by Minister Colbert has in fact launched a new methodological path in naval design, starting a virtuous path that in the space of a century will lead the craftsmanship of the project and shipbuilding to a system of "industrialization" where the shipwright will no longer be the only architect of the construction but will become part of a staff of technicians who will see involved in the project and shipbuilding, naval architects and engineers, shipyard technicians such as carpenters, blacksmiths, caulkers, etc., in addition to the artists [27].

Figure 4: Plate n. 47, longitudinal section of a vessel.

In the contemporary literature can be observed two distinct ways of narrating the construction of vessels; a more descriptive one, where the ship is "told" in written form practically without images and with only summary didactic illustrations, and a more encyclopaedic one, where the ship is described and illustrated in terms of construction elements. Compared to them, the *Album de Colbert* is instead a manual for the construction of a vessel, step by step, almost a contemporary "instruction booklet" for assembly, the archetype of the industrialization processes of these last thirty years. The *Album* is the representation of a production process that will see its complete and exhaustive synthesis in the formation of the construction plans in the following centuries with the drafting of general plans, water lines, significant sections, and the development of decks, in consequence of the progress of mathematics and descriptive geometry thanks to Gaspard Monge (1746 -

1818), who was also appointed Minister of the French Navy in 1792, and of his followers including Jean-Victor Poncelet (1788 - 1867), and of the representation through the technical design.

The *Album de Colbert* is a theatrical "scene" where the ship is the central character, a mime who tells her story, showing herself scene after scene, through the objects that compose it and through the images that describe it. Her existence is made real only by representation, and thanks to man's creative capacities of imagination, it materializes even in the absence of the object itself. In this sense, the *Album de Colbert* stands as an element of conjunction between a technical-practical knowledge and a technical-scientific knowledge, where illustration and representation are the distinctive tools of that junction that will see the development of Naval Architecture and Naval Science beyond the empirical knowledge of the construction tradition in the naval field.

Figure 5: Plate n. 50 showing 80 guns vessel fully rigged, fully fitted out, and ready to get underway.

6. ACKNOWLEDGEMENTS

This publication was written in "*Art of building and aesthetics in naval iconography from the Middle Ages to the Modern Age*" research area and it was financed with Department of Architecture and Design 'PRA 2019' funds of the University of Genoa.

7. REFERENCES

1. *Dessins représentant toutes les phases de la construction d'un vaisseau, dit «Album Colbert»* [See Paper N. SH 140, XVIIe siècle. Bibliothèque Vincennes].
2. VERGÉ-FRANCESCHI, M., RIETH, É., *France maritime au temps de Louis XIV: édition critique des deux Albums dits de Jouve et de l'Album de Colbert*. Paris: Ed. du Layeur, 2001. The work reproduces three collections of engravings dedicated to the French Navy and made in the seventeenth century on the initiative of Colbert. One, known as the Album de Colbert, is dedicated to building a warship. The other two, known as the *Albums de Jouve. Desseins de tous les bâtiments qui naviguent sur la Méditerranée, par Jean Jouve, de Marseille* (1679), probably engraved in 1679 and published in 1710, describe the fishing and trading vessels.
3. DE LA RONCIÈRE, C., *Un tricentenaire : un grand ministre de la marine, Colbert*. Paris : Plon-Nourrit et Cie, 1919.
4. ALLARD, C., *L'art de batir les vaisseaux, et d'en perfectionner la construction ; de les garnir de leurs apparaux, les mettre en funin, les manœuvrer, &c. ... outre, Les pavillons de divers etats. Le tout tiré des meilleurs Auteurs Hollandois, comme Witsen, van Eyk, Allard, &c.* Amsterdam: David Mortier, 1719.
5. WITSEN, N., *Architectura navalis et regimen nauticum Ofte Aaloude en Hedendaagsche Scheeps-bouw en bestier*. Amsterdam: Pieter en Joan Blaeu, 1690, second edition (First edition, Amsterdam: Casparus Commelijn; Broer en Jan Appelaer, 1671).
6. VAN IK, C., *De Nederlandsche Scheeps-Bouw-Konst Open Gestelt: vertoonende Naar wat Regel, of Evenredenheyd, in Nederland meest alle Scheepen werden gebouwd; mitsgaders Masten, Zeylen, Ankers, en Touwen, enz. daar aan gepast.* Delft: Andries Voorstad; Amsterdam: Jan ten Hoorn, 1697.
7. BOISSONNADE, M.P., 'La marine marchande, le port et les armateurs de La Rochelle à l'époque de Colbert (1662-1683). Comité des travaux historiques'. *Bulletin de la section de géographie,* t. XXXVII, 1922; Paris: Imprimerie nationale, 1923; p. 1-45 (Cited in: *Bulletin de la Section de géographie*, Tome XXXVI, Année 1921. Paris: Imprimerie nationale 1922; p. XXII-XXIII).
8. ACERRA, M., MERINO, J., MEYER, J.; VERGÉ-FRANCESCHI, M., *Les marines de guerre européennes: XVIIe-XVIIIe siècles*. Paris: Presses de l'Université de Paris-Sorbonne, 1998 (deuxième édition).
9. VERGÉ-FRANCESCHI, M., RIETH, É., *La France maritime au temps de Louis XIV: édition critique des deux albums dits de Jouve et l'album de Colbert*. Paris : Ed. du Layeur, 2001.
10. DUPONT, A., FARDET, M., *L'Arsenal de Colbert Rochefort.* Rochefort : Centre Internationale de la Mer, 1986.
11. BLANCHARD, A., *Les Ingénieurs du roy, de Louis XIV à Louis XVI : étude du corps des fortifications*. Montpellier : Impr. Déhan, 1979.

12. ACERRA, M., *Rochefort et la construction navale française*. Paris : Librairie de l'Inde éd., 1993; PIERRE, J., *Les constructions navales à Rochefort*. Paris : le Croît vif, 2003.
13. PÈNE and CASSINI, J.-D., *Le Neptune françois ou Atlas nouveau des cartes marines levées et gravées par ordre exprès du roy. Pour l'usage de ses armées de mer ... Revue & mis en ordre par les Sieurs Pene, Cassini & autres*. Paris : H. Jaillot, 1693. FRÉMONT D'ABLANCOURT, N.; MORTIER, P., CHAZELLES, J.-M. de, *Suite du Neptune françois; ou atlas nouveau des cartes marines ... Levées par ordre exprès des roys de Portugal... et données au Public par... Monsieur d'Ablancourt, dans lequel on voit la description exacte de toutes les côtes du Monde...* Amsterdam: Pierre Mortier, 1700.
14. BIERRE, C., 'Jean-Baptiste Colbert, Bâtisseur de la Nation ou La Politique du Grand Dessein', in *L'Europe, vecteur d'une reprise de l'économie mondiale*, edited by TENNENBAUM, J., LAROUCHE, L.H., BIERRE, C. Paris: Institut Schiller, 1992.
15. ACERRA, M., 'Les constructeurs de la marine (XVII e -XVIII e siècle)', *Revue Historique*, Avril-Juin 1985, T. 273, Fasc. 2 (554) (Avril-Juin 1985), p. 283-304.
16. DE LA RONCIERE, C., *Un tricentenaire : un grand ministre de la marine, Colbert.* Paris : Plon-Nourrit et Cie, 1919 ; p. 158 and SERIS, J.-P., *Machine et communication: du théâtre des machines à la mécanique industrielle*. Paris : J. Vrin, 1987 ; p. 69.
17. VASLIN, J.-M., *Les chênes français, le "trésor" de Colbert.* Le Monde, 27 avril 2011.
18. GALLON, DE (officier de la maîtrise des eaux et forêts de Rochefort). *Conférence de l'Ordonnance de Louis XIV. Du mois d'Août 1669. Sur le fait des eaux et forêts, Avec les Édits, Déclarations, Coutumes; ... depuis l'an 1115 jusqu'à présent. Contenant les loix forestières de France. Nouvelle édition, ...* 2 vol. Paris : Bernard Brunet, 1752.
19. DOBRENKO, A.; PALMER, A., 'The model of La Légère in the Central Naval Museum, St Petersburg: master shipbuilder Blaise Pangalo in Peter the Great's shipyards', *The Mariner's Mirror*, 86, 2000 - Issue 1, p. 37-49.
20. WINFIELD, R.; ROBERTS, S.S., *French Warships in the Age of Sail 1626–1786. Design, Construction, Careers and Fates*. Barnsley: Seaforth Publishing, 2017.
21. GOUJON, L.-J.-M.A., *Des Bois propres aux Constructions Navales. Manuel à l'usage des Agens Forestiers et Maritimes. Contenant les Lois, Règlemens et Instructions relatifs à la disposition et à l'usage des bois dits de Marine. Accompagné de vingt-sept figures gravées et enluminées, réduites sur celles données par l'Administration générale des Forêts, à la suite des instructions officielles du 20 Messidor XI. Suivi d'un Dictionnaire des principaux termes d'Architecture Navale.* Paris: chez Goujon fils, Imprimeur-Libraire, an XI, 1803.
22. *Dessins représentant toutes les phases de la construction d'un vaisseau, depuis sa mise en chantier jusqu'à son lancement et son arrimage* [XVIIe siècle. 50 planches. SH 140 ex 1513, Bibliothèque Vincennes]. In the same archive there is a reduced edition copy: *Dessins représentant toutes les phases de la construction d'un vaisseau, depuis sa mise en chantier jusqu'à son lancement et son arrimage. Parmi ces beaux dessins, on peut signaler des vues des différentes poupes de vaisseaux royaux* [XVIIe siècle. 40 planches. SH 141 ex 1514, Bibliothèque Vincennes].
23. WINFIELD, R., ROBERTS, S.S., *French Warships in the Age of Sail, 1626–1786: Design, Construction, Careers and Fates*. Barnsley: Seaforth Publishing, 2017; p. 56-60.
24. *Ordonnance de Louis XIV pour les armées navales & arsenaux de marine. Du 15 avril 1689*. Paris: Prault Pere, 1764. Livre treizième. "De la construction des vaisseaux. Titre II. De la construction et proportions des vaisseaux. Art. I".
25. RIETH, E., *Le maître-gabarit, la tablette et le trébuchet : Essai sur la conception non-graphique des carènes du Moyen âge au XXe siècle*. Paris: CTHS, coll. Histoire des sciences, 1996.
26. VIAL-DU CLAIRBOIS. *Dictionnaire Encyclopédique de Marine.* Tome Premier. Paris: Panckoucke, 1793; Article second: Méthode pour tracer le plan de projection d'un vaisseau de 70 canons, pp. 516-533, see p. 520.
27. See TACCHELLA, C., *The Royal Louis (1668), A Sailing Masterpiece*, Pending publication.

8. AUTHORS BIOGRAPHY

Massimo Corradi is Associate Professor of History of Science and Structural Mechanics at the Polytechnic School of the University of Genoa. Visiting professor and invited lecturer: Université Catholique de Louvain La Neuve (Belgium), Open University of London, Universidad Nacional Autónoma de México, Université de St. Etienne (France), Universidade de Brasilia (Brasil), School of Art and Architecture, Shiraz University (Iran), Art University of Isfahan (Iran) and Iranian Architecture Center, Tehran (Iran). Invited lecturer in several International Conferences on the subject of History of Science and Structural Mechanics, and the relationship between Mechanics and Architecture, he is author of about one hundred publications on the subjects of History of Science and Structural Mechanics, and History of naval science and shipbuilding.

STRUCTURAL ASSESSMENT AND SCANTLINGS OF TRADITIONAL SMALL CRAFTS

J.-B. R. G. Souppez, Aston University, UK

SUMMARY

The scantlings of traditional wooden crafts often originate from the extensive experience of designers and boatbuilders. With the contemporary regain of interest for historical replicas and timber construction, and the ever-critical necessity to minimise displacement, a compromise between the original scantlings and modern structural assessment must be struck. The latter heavily relies on rules-based design, driven by formal regulations, though a number of empirical methods also exist. Consequently, to assess the comparative structural requirements between traditional small crafts, empirical methods and regulatory requirements, case studies will be undertaken on small crafts. Ultimately, the results showcase the differences between original specifications, the recommendation of simplified methods, and modern rules, with the latter allowing for weight savings.

1. INTRODUCTION

Small crafts have heavily influenced the development of humanity, with evidence dating back to as early as the 6th millennium BC [1]. Throughout the vast majority of this time, the design and construction were unregulated, and relied on trial and error. Some of these errors remain particularly famous, such as the tragic capsize and sinking of the Swedish warship Vasa [2] in 1628, highlighting the vital need for sufficient stability and lowering the vertical centre of gravity.

The concept of pleasure craft is very recent, with the first instances being recorded during the 19th century, ahead of the two waves of development experienced in the 20th century [3] by the recreational small craft industry.

Firstly, the 1930s benefited from the fast improvements in engine power, lower weight and more accessible cost, following World War I and the progresses made in the automotive industry. Furthermore, as all pleasure industries, small crafts were made more popular thanks to the introduction of paid holiday throughout Europe. At this time, vessels remain small (8/9 m), primarily wooden runabouts, and able to reach 30 knots [3].

Later on, the 1960s witnessed the second wave of development in small crafts. Thanks to the emerging use of composites, large scale production increases, with bigger and faster vessels, now featuring accommodation. Once again, this was supported by the social developments and longer paid holiday across Europe.

Still, small leisure crafts remained unregulated. Larger vessels had seen the birth of SOLAS (Safety Of Life At Sea) in 1914 [4], following the sinking of the Titanic in 1912 [5]. But for small vessels, under 24 m in length, this would not come until the aftermath of the 1979 Fastnet Race [6], where the fleet was hit by a violent storm [7], leading to the loss of 15 yachtsmen, 75 capsizes and 5 sinkings. This triggering event highlighted the vital need for strict regulation on small pleasure crafts, including the stability and structure of vessels.

In time, this led to the Recreational Craft Directive, originally approved in 1994 [8], and prompted the development of numerous supporting standards, including the ISO 12217 [9] for stability, and the ISO 12215 [10] for scantlings.

More recently, the revised Recreational Craft Directive (RCD II) [11] resulted in renewed supporting ISO standards. Of particular interest to this paper are the developments made in the structural regulations [12], covering sailing [13], power [14] and commercial crafts [15]. Despite some limitation for modern wooden boats [16], historical crafts fall under a special regime.

Indeed, as stated in the RCD II [11]:

"This Directive shall not apply to [...] original historical watercraft and individual replicas thereof designed before 1950, built predominantly with the original materials and labelled as such by the manufacturer."

In this instance, the compliance is assumed on the basis of the success of historical designs. However, with the regain of interest for modern replicas [17], new designs inspired by historical ones [18], and the application of modern yacht design techniques to historic crafts [19], the compatibility of historical scantlings with modern rules and regulations has become of interest. Consequently, this paper will investigate the compliance of scantlings from historical crafts in light of the latest structural regulations, namely the ISO 12215-5:2019 [20], through the case studies of two small vessels. In additional, empirical methods for wooden boat scantlings will also be applied for comparison purposes.

2. METHODS

Three methods will be employed for scantlings determination; two empirical ones, and a formal regulation.

First, the method proposed by Gerr [21] will be applied. Derived for the study of existing vessels, it offers a very simple approach to structural sizes for small crafts, with each component being related to a function of the overall size of the vessel, termed scantling number, Sn, and defined as:

$$Sn_{Gerr} = \frac{LOA \times BOA \times D}{28.32} \quad (1)$$

Where:

LOA	Length overall (m)
BOA	Beam overall (m)
D	Depth of hull (m)

Then, the method developed by MacNaughton [22] will be featured. It also expresses the required size of structural components as a function of a single input, related to the size of the vessel, also termed scantling numeral, but this time defined as the cube root of the displacement of the vessel (in imperial units):

$$Sn_{MacNaughton} = \sqrt[3]{\nabla} \quad (2)$$

Where:

∇	Displacement (ft^3)

Lastly, the ISO 12215-5:2019 [20] will be applied. It is to be noted that the standard is only recognized for modern timber construction (namely strip planking, cold moulded, or a combination of both, as well as plywood), but not traditional techniques, such as carvel planking.

However, the method will be applied in this instance, as the underpinning theory remain very simple (namely assumption that a panel may be approximated as a built-in be under uniformly distributed load), despite being intended for modern construction. It should also be noted that it has been suggested cold moulded structures would be best analysed in a similar manner as composite laminates [23], i.e. in a ply-by-ply analysis, as opposed to a quasi-isotropic one.

3. CASE STUDIES

3.1 THAMES A RATER

The historical Thames A Rater class saw its birth in the late 19th century [24], and while the vessels still feature the same original wooden hulls [25], significant technological advances have been made over the years to keep the class competitive [26]. This led to a number of recent studies [19, 27, 28] focussed on performance optimisation, as well as characterisation of original designs, such as *Scamp*, a 1902 Linton Hope design, reproduced from the original linesplan [29], and presented in Figure 1.

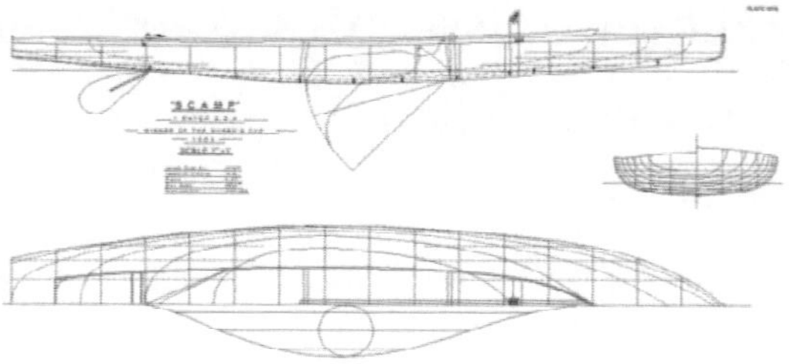

Figure 1: *Replica of the Scamp linesplan [29].*

The modern developments in rig and sail designs, added onto the original timber hulls, can be seen in Figure 2, comparing the A Rater *Vagabond* over 100 years apart.

Figure 2: *Rig in 1907 (left) [28] and 2014 (right) [28].*

3.2 DARK HARBOR 17.5

Designed in 1908 by B. B. Crowninshield [30], the Dark Harbor 17.5 is a traditional day sailor, with well documented plans and scantlings [31], and depicted in Figure 3 and Figure 4. The design has also benefited from recent interest with the aim of developing a modern replica [17, 32], shown in Figure 5, better suited to today's market, while also incorporating a more advanced construction method.

As such, it would no longer fall under the pre-1950 exemption in terms of regulatory compliance, hence the interest in the required scantlings for this vessel under current rules.

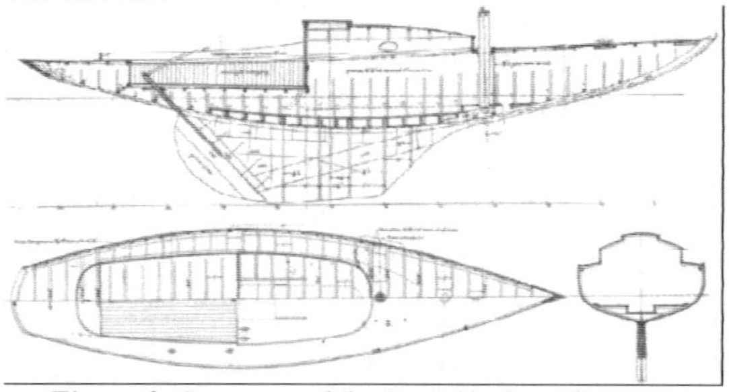

Figure 3: *Structure of the Dark Harbor 17.5 [31].*

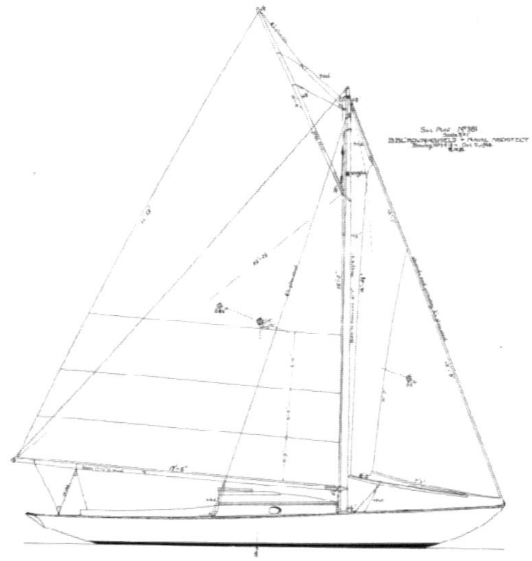

Figure 4: *Sail plan of the Dark Harbor 17.5 [31].*

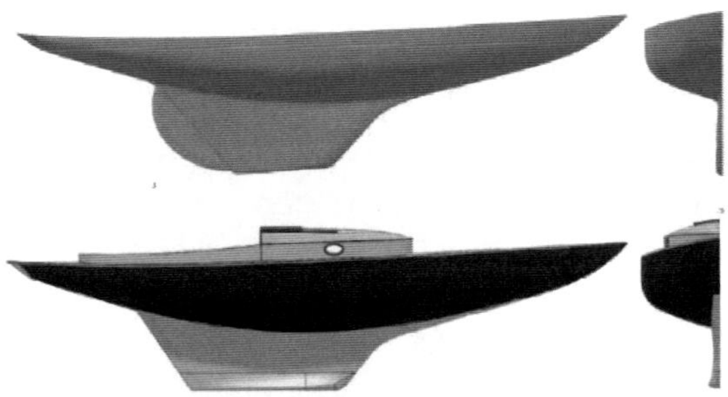

Figure 5: *Original Dark Harbor 17.5 (top) [31] and modern replica (bottom) [32].*

3.4 HYDROSTATICS AND SCANTLINGS

For the case studies under consideration, the main hydrostatics, together with the resulting scantling numbers (in accordance with both Gerr's [21] and MacNaughton's [22] definitions) are presented in Table 1.

It should be noted that, although both labelled 'scantling numbers', there is no relationship between these two quantities across both empirical methods. Consequently, the quantitative differences in their values are of no importance at this stage.

Table 1: *Main hydrostatics and scantlings numbers.*

Parameter	A Rater (Scamp)	Dark Harbor 17.5
LOA (m)	8.28	7.92
Lwl (m)	5.17	5.34
BOA (m)	1.90	1.91
Bwl (m)	1.64	1.84
D	0.47	0.86
∇ (m^3)	0.548	1.513
∇ (ft^3)	19.35	53.43
SN_{Gerr}	0.26	0.46
$SN_{MacNaughton}$	6.45	17.81

4. RESULTS

Empirical methods typically suffer from a number of limitations; these include:

- An absence of consideration for the timber species considered, and therefore the mechanical properties of the material.
- There is no allowance for the design pressure applied to the craft, which would vary with the vessel's speed, but also operating profile (eg: inland versus offshore craft).
- The actual geometry of the panel, in terms of its size and curvature, are not accounted for.

There are therefore flaws in empirical methods, which do however benefit from a level of simplicity far greater than that of regulatory requirements, and thus often found their use restricted to the very early stages of the design. Furthermore, no technical background is provided for these methods. Ultimately, this yields very inconsistent results, that widely diverge from the actual scantlings of the vessels, as presented in Table 2.

Table 2: *Comparison of the main scantlings.*

Method	Scantling	A Rater (Scamp)	Dark Harbor 17.5
Gerr	Planking thickness (mm)	11.0	13.8
	Square frame (mm)	20.3	25.5
	Frame spacing (mm)	179.2	208.7
Mac Naughton	Planking thickness (mm)	24.6	67.9
	Square frame (mm)	41.0	113.1
	Frame spacing (mm)	245.7	678.6
Actual	Planking thickness (mm)	7.9	19.1
	Square frame (mm)	12.7	22.2
	Frame spacing (mm)	152.4	203.2

In comparison, the ISO 12215-5 [20] addresses the major limitations of the empirical methods. Firstly, the planking thickness is derived from a simplified structural analysis, assuming the panel as a built-in beam under uniformly distributed load, thereby providing some technical background. The plating thickness under this small craft regulation is given by Equation 3:

$$t = b \times k_c \times \sqrt{\frac{P \times k_{2b}}{1000 \times \sigma_d}} \qquad (3)$$

Where:

t	Thickness (mm)
b	Panel's short side (mm)
k_c	Curvature coefficient (-)
P	Pressure (kN/m^2)
k_{2b}	Aspect ratio correction factor (-)
σ_d	Design stress (N/mm^2)

While more complex in its formulation, and requiring a higher number of inputs, this equation accounts for:

- The panel's geometry, in terms of its physical size (b), curvature (k_c) in both the longitudinal and transversal directions, and aspect ratio (k_{2b}).
- The pressure (P), which varies with multiple factors, including the design category of the vessel (inland, inshore, offshore and ocean), and the longitudinal position along the length of the vessel.
- The mechanical properties of the actual timber species considered, and building on a wealth of previous research for various timbers [33, 34, 35, 36, 37, 38].

While a strict comparison cannot be undertaken between all methods, the application of regulatory rules to traditional craft consistently yields far lower structural requirements, whether for the planking or framing. This is supported by the number of recent investigations into modern replicas and novel structural designs, that routinely achieved structural weight savings [17, 19, 28].

Despite the impossibility to generate like-for-like comparison between the actual scantlings, those advised by empirical methods, and those required by modern regulations, very clear results were yielded. With too many limitations, empirical method over-structure historical crafts (which may already be deemed over-structured by today's standards), particularly small inland/inshore ones. Conversely, vast weight savings can be achieved under the ISO 12215-5, which should therefore be considered, where applicable, even for historical crafts that may be exempt from regulatory compliance due to their design dating back prior to 1950 and constructed primarily as per the original vessel specifications.

5. CONCLUSIONS

This paper used two historical crafts, namely the Thames A Rater and the Dark Harbor 17.5, as case studies to comparatively assess the relevance of empirical scantlings methods and contemporary rules and regulations.

With vastly higher structural requirements, empirical methods may find there place in very early design stages, but did not prove suitable to determine the required size of structural components. These methods suffer from their simplicity, and do not allow to capture all the necessary aspects of a craft and its operation to yield relevant scantlings.

On the other hand, thanks to the lower structural requirements of the ISO 12215-5, weight savings could be achieved on historical crafts. This is particularly important, as the fit out of these vessels in modern days tends to be far heavier than originally, due to the addition of engine, batteries, life-saving equipment, etc…

Remembering that crafts designed before 1950s and built in accordance with the original scantlings are exempt of regulatory compliance, this paper demonstrates that there is a strong benefit in complying with modern standards. Consequently, it would be advised to investigate this option, which may also provide further consumer confidence in the reliability of the vessel.

6. REFERENCES

[1] CRTER, R., 'Boat Remains and Maritime Trade in the Persian Gulf During the Sixth and Fifth Millennia BC', *Antiquity*, vol 80(307), pp 52-63, 2006.

[2] KESSLER, E. H., BIERLY, P. E & GOPALAKRISHNAN, S., 'Vasa Syndrome: Insights from a 17th-Century New-Product Disaster', *Academy of Management Perspectives*, vol 15, no 3, 2001.

[3] ISSC COMMITTEE V.8, 'Yacht Design', *Proceedings of the 18th International Ship and Offshore Structures Congress*, 2012.

[4] WHEELER, E. P., 'International Conference of Safety of Life at Sea', *The American Journal of International Law*, vol 8, no 4, 1914.

[5] CONRAD, J. 'Some Reflexions, Seamanlike and Otherwise, on the Loss of the Titanic', *The English Review 1908-1937*, vol 11, 1912.

[6] O'DONNERL, B., 'The Fastnet Race 1979', *British Medical Journal*, vol. 281, 1980.

[7] ADLARD COLES, K., 'The Fastnet Storm of 1979', *The Journal of Navigation*, vol 33, issue 2, 1980.

[8] EUROPEAN PARLIAMENT, 'Directive 1994/25/EC on Recreational Craft', *Official Journal of the European Union*, 1994.

[9] INTERNATIONAL ORGANISATION FOR STANDARDIZATION, 'BS EN ISO 12217 - Small Craft - Stability and Bouyancy Assessment and Categorization', *International Organisation for Standardization*, 2015.

[10] INTERNATIONAL ORGANISATION FOR STANDARDIZATION, 'BS EN ISO 12215 - Small Craft - Hull Construction and Scantlings', *International Organisation for Standardization*, 2019.

[11] EUROPEAN PARLIAMENT, 'Directive 2013/53/EU on Recreational Craft and Personal Watercraft', *Official Journal of the European Union*, 2013.

[12] SOUPPEZ, J.-B. R. G. & RIDLEY, J., 'The Revisions of the BS EN ISO 12215', *Marine Sector Showcase* - Composite UK, Southampton, UK, 2017.

[13] SOUPPEZ, J.-B. R. G., 'Structural Design of High Performance Composite Sailing Yachts under the New BS EN ISO 12215-5', *Journal of Sailing Technology*, issue 2, pp 1-18, 2018.

[14] SOUPPEZ, J.-B. R. G., 'Designing the Next Generation of Small Pleasure and Commercial Powerboats with the Latest ISO 12215-5 for Hull Construction and Scantlings', *1st SNAME / IBEX Symposium*, Tampa, Florida, US, 2019.

[15] SOUPPEZ, J.-B. R. G., 'Structural Analysis of Composite Search and Rescue Vessels under the New BS EN ISO 12215-5'. *SURV9 - Surveillance, Search and Rescue Craft*, The Royal Institution of Naval Architects, London, UK, pp 1-4, 2018.

[16] SOUPPEZ, J.-B. R. G., 'Timber Construction: an Experimental Assessment of the Strength of Scarf Joints and the Effectiveness of Various Adhesives for Laminated Wood', *Historic Ships, The Royal Institution of Naval Architects*, London, UK, 2020.

[17] ALESSIO, L. G., SOUPPEZ, J.-B. R. G. & HAGE, A., 'Design Evaluation and Alteration of the Dark Harbor 17.5: Case Study of a Modern Replica', *Historic Ships, The Royal Institution of Naval Architects*, pp 121-127, London, UK, 2016.

[18] LINDEN, V. & SOUPPEZ, J.-B. R. G., 'Sailing towards Sustainable Trading with Wooden Cargo Schooner', In the Proceedings of the *British Conference of Undergraduate Research*, Sheffield, 2018.

[19] SOUPPEZ, J. B. R. G., 'On the Applications of Modern Naval Architecture Techniques to Historical Crafts', *Historic Ships, The Royal Institution of Naval Architects*, pp 1-15, London, UK, 2016.

[20] INTERNATIONAL ORGANISATION FOR STANDARDIZATION, 'BS EN ISO 12215-5:2019 - Small Craft - Hull Construction and Scantlings. Part 5: Design Pressures for Monohulls, Design Stresses, Scantlings Determination', *International Organisation for Standardization*, 2019.

[21] GERR, D., 'The Elements of Boat Strength', *International Marine, McGraw-Hill*, 1999.

[22] MACNAUGHTON YACHT DEIGNS, 'Scantling Rules', *MacNaughton Yacht Designs*, 2010.

[23] LOSCOMBE, R., 'Structural Design Considerations for Laminated Wood Yachts', *International Conference on the Modern Yacht*, Portsmouth, UK, 1998.

[24] LAITY, M., 'Thames Clubs', *Classic Boats*, 2001.

[25] LAITY, M., 'Bell Boy', *Classic Boats*, 2007.

[26] THAMES SAILING CLUB, 'The Thames A Rater Association Rules', *Thames Sailing Club*, 2007.

[27] SOUPPEZ, J.-B. R. G., 'Design and Production of a Wooden Thames A-Rater Class Sailing Yacht', *Master Thesis, The University of Auckland*, 2015.

[28] THOMAS, J. & SOUPPEZ, J.-B. R. G. 'Comparative Performance Prediction of Historical Thames A Rater Class Designs', *Historic Ships, The Royal Institution of Naval Architects*, pp 67-76, London, UK, 2018.

[29] KEMP, D., 'Dixon Kemp's Manual of Yacht and Boat Sailing and Yacht Architecture. New and Eleventh Edition. Edited by B. Heckstall-Smith and Linton Hope', *Horace Cox, London*, 1913.

[30] CROWNINSHIELD, B. B., 'Dark Harbor 17½', *Dark Harbor Design Plan,* 1908.

[31] WOODEN BOATS, '25'10" Sloop, Dark Harbour 17½', *Wooden Boats*, undated.

[32] ALESSIO, L. G., 'Redesign of a Classic Sailboat with FEA Investigation of the Plate Curvature', *Master Thesis*, 2017.

[33] US GOVERNMENT, 'Design of Wood Aircraft Structure,'A*nc-18 Bulletin,* 1951

[34] STARTLING PUBLISHING CO, 'Encyclopedia of Wood', *Startling Publishing Co*, 1989.

[35] PRINCESS RISBOROUGH LABORATORY, 'The Strength Properties of Timber in Metric Units', *Building Research Establishment*, 1978.

[36] BRITISH STANDARDS, 'BS EN 408:1995, Timber Structures - Structural Timber and Glued Laminated Timber - Determination of Some Physical and Mechanical Properties', *British Standards*, 1995.

[37] BRITISH STANDARDS, 'BS EN 1995-1-1:2004, EUROCODE 5, Design of Timber Structures, Plus TRADA Guidance Document N° 3', *British Standards*, 2004.

[38] DENSCH, H. E. & DINWOODIE, J. M., 'Timber - Structure, Properties, Conversion and Use', 7th edition, *MacMillan Presss Ltd*, 1996.

7. AUTHORS BIOGRAPHY

Jean-Baptiste R. G. Souppez is the Senior Teaching Fellow in Mechanical Engineering and Design at Aston University. As the UK Principal Expert in Small Craft Structures, he is in charge of representing the interest of the British marine industry in the development of international structural regulations (ISO12215) while also overseeing all small craft regulatory developments on behalf of the Institute of Marine Science, Engineering and Technology (IMarEST). He is also the Deputy Editor-in-Chief of the Society of Naval Architects & Marine Engineers (SNAME) Journal of Sailing Technology. His research in fluid dynamics features a wide range of experimental methods supporting his cutting-edge expertise in hydrofoils and award-winning work on the aerodynamics of downwind racing yachts sails. He contributes to the international Erasmus Mundus Master in Advanced Design in Ship and Offshore Structures (EMship+) as a Visiting Professor and Research Supervisor at the University of Liege, while promoting professional development as an Associate Tutor for the MLA College, supervising post-graduate researchers and enabling established professionals to achieve Chartered Engineer (CEng) status through the IMarEST.

TIMBER CONSTRUCTION: AN EXPERIMENTAL ASSESSMENT OF THE STRENGTH OF SCARF JOINTS AND THE EFFECTIVENESS OF VARIOUS ADHESIVES FOR LAMINATED WOOD

J.-B. R. G. Souppez, Aston University, UK

SUMMARY

Timber construction has recently seen a significant regain of interest across a range of industries, owing to contemporary concerns for sustainability. In the marine industry, historic principles of traditional wooden boatbuilding remain present, with empirical rules still common practice, as is the case for scarf joints. Moreover, laminated wood is made more attractive and efficient thanks to modern adhesives. However, with the progresses made in structural analysis, these assemblies can now be refined based on scientifically informed evidence. Consequently, this paper will employ destructive testing to tackle two distinct cases. On the one hand, the strength of feathered (plain) scarf joints as a function of their slope will be evaluated. On the other hand, the effectiveness of a range of adhesives will be ascertained for the purpose of laminated manufacturing. Ultimately, the results will be compared to both the strength of solid wood and the mechanical properties assumed by modern scantling regulations, revealing significant differences. The research findings provide a better understanding of these fundamental timber construction principles, supporting designers and builders alike in making informed choices and promoting safer regulatory compliance. It is also anticipated these findings will impact structural design beyond the wooden boatbuilding field, with applications in sustainable buildings and architecture.

1. INTRODUCTION

Traditional timber construction has significantly impacted maritime transportation and yacht design, and some historical principles remain key elements of today's wooden boatbuilding. Despite the wealth of experience that originates from trial and error, limited scientific background dedicated to wooden boats exist. Consequently, this paper tackled two areas of particular interest, namely scarf joints and laminated timber, using destructive testing to quantify the mechanical properties. These are compared to experimental values for solid timber as well as the allowable regulatory properties. The aim is to provide scientific evidence to better support the use of timber construction in modern yacht design and construction. The experiments and analyses inherent to scarf joints are discussed in Section 2, while laminated timbers are discussed in Section 3. Ultimately, Section 4 summarises the findings and provides recommendations regarding the mechanical properties for safer and more reliable wooden yachts.

2. SCARF JOINTS

2.1 BACKGROUND

The use of scarf joints is a fundamental part of timber construction and traditional boatbuilding, allowing to overcome the natural restrictions in sizes to achieve components as large as necessary. Today, the use of scarf joints also extends into composites structures [1]. Nevertheless, modern advances in structural engineering leading to progressively lighter boats, coupled with significant progress in adhesives [2] and the contemporary regain of interest for wood as an engineering material [3; 4] call for a new understanding of the mechanical properties and strength of scarf joints compared to solid timber.

Scarf joints are characterised with a length-to-thickness ratio, and are further categorised by their various types, as depicted in Figure 1.

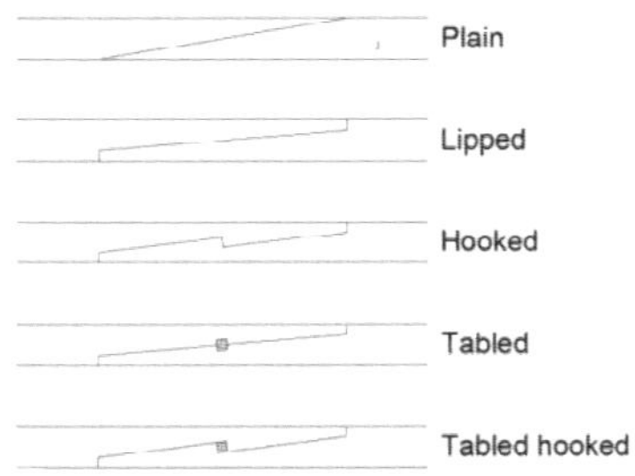

Figure 1: *Examples of scarf joints [5].*

While more complex joints have received more attention due to their use in civil engineering [6], plain scarfs remain the prevalent option in traditional boatbuilding, and thus will be the primary focus of this investigation.

Historically, scarf ratios have been driven by the location of the scarf on the vessel: 4:1 for planks, 6:1 (possibly 8:1) for keels, and 12:1 for spars. The Lloyd's rules [7] published in 1979, although no longer applicable, stated that plank scarfs should not have a length-to-thickness ratio less than 4 (rule 4707), adjacent planks shall not have scarfs within 1.2 m of each other, and a minimum of three complete planks shall separate scarfs in the same transverse plane. In addition, keel scarfs shall have a ratio no less than 6:1 (rule 4302), and the keel and hog scarfs should be spaced by at least 1.5 m (rule 4303), while being clear of engine bearers and masststeps. In those historical instances, it can be deduced that an increased scarf ratio leads to greater strength, though scarfs still represent weak spots that should be spaced out and not subjected to highly localized loads.

Very few instances of guidelines regarding the effectiveness of scarfs for boatbuilding applications are present in the literature. Birmingham [8] suggests that the efficiency of scarfs ranges from 65 percent of the strength of solid timber for a 4:1 ratio and up to 95 percent for a 20:1 ratio. On the other hand, Gerr [9] recommends a 12:1 ratio that will achieve 90 percent of the strength of solid timber. Lastly, an 8:1 ratio is advised for greater strength, with a 12:1 ratio being recommended for spars [10], with additional rules of thumbs regarding spacing and slope suggested [11]. The origin of these various values is however not clear, nor is their accuracy when utilizing different glues and wood species, and no underpinning scientific data is presented to support the claims made.

Consequently, in order to provide a detailed analysis of how scarf ratios affect the strength of timber components, destructive structural testing was undertaken on European oak samples (*Quercus spp*, having a density no less than 690 kg.m^{-3} at 12% moisture content), joined together with feathered scarfs glued with epoxy. In order to faithfully replicate a typical boatyard scenario, the samples were manufactured from different quarter sawn boards, never joining samples from the same board. Ratios of 4:1, 8:1, 12:1, 16:1, and 20:1 were tested, comparing all of them to solid samples. The aim was to ascertain the relative strengths of the various scarf ratios, for the wood species and adhesive utilised in this instance, both prominent in the boatbuilding industry, to provide designers and builders with relevant and reliable information; and an evaluation of regulatory default values.

2.2 EXPERIMENTAL TESTING

2.2 (a) Manufacturing

The samples sizes were 400 mm long, by 20 mm wide, by 20 mm thick, with in excess of 5 samples per tested configuration, in accordance with the relevant standard [12]. An example of joints prior to gluing is shown in Figure 2, with the adhesion process being conducted under camping pressure for the duration of the epoxy's cure.

Figure 2: *Samples machined prior to gluing [4].*

2.2 (b) Method

To assess the strength and mechanical properties of samples, a number of destructive tests can be employed; for timber, four-point bending is preferred, as it allows one to establish the ultimate flexural strength and elastic modulus (also known as Young's modulus). In four-point bending tests, the sample is simply supported at each extremity, while the load is applied evenly at two locations equidistant from the center, as depicted in Figure 3. In this instance, all tests were conducted on a Lloyds Instruments LR 30k tensile machine. To ensure the reliability, accuracy, and repeatability of the results obtained, the BS EN 408 standard [12] was applies.

Figure 3: *Experimental setup [4].*

2.3 QUALITATIVE RESULTS

From the experimental testing, typical load-deflection curves for the various scarf ratios compared to solid timber were obtained, see Figure 4, yielding three main findings.

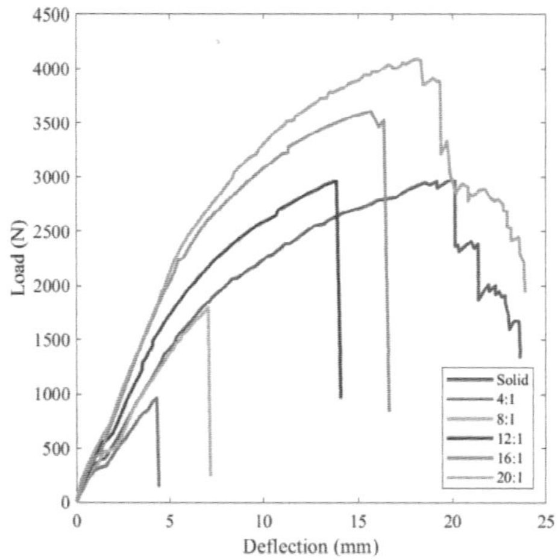

Figure 4: *Typical load-deflection curves [4].*

First, the results clearly demonstrate that the small scarf ratios (4:1 and 8:1) have a lesser resistance compared to solid timber, whereas the opposite is true for the higher ratios (12:1, 16:1, 20:1). Practically, this means that higher scarf ratios have a higher modulus of elasticity than the solid timber and will be able to carry more load for a given deflection.

The second important result is that solid timber withstands the most deflection before ultimate failure; in other words, despite not carrying as much load as a high-ratio scarf, solid timber is able to deflect much farther than scarfed samples.

Finally, there is an interesting shift in the failure mechanisms, shown in the fracture behavior. For ratios ranging from 4:1 to 16:1, the fracture is sudden and abrupt, with no strength left. In these cases, it was the epoxy bond that failed. Conversely, solid timber and the 20:1 scarf do retain some strength, as it is the timber and not the adhesive that failed in those instances. The two failure modes are presented in Figure 5. This proved true for all samples tested with the exception of a single 16:1 ratio where a combination of timber and glue failure (attributed to a weak spot in the timber) was noticed. It is to be noted that these findings will be affected by the adhesive employed, and could vary for alternative glues, as well as timber species.

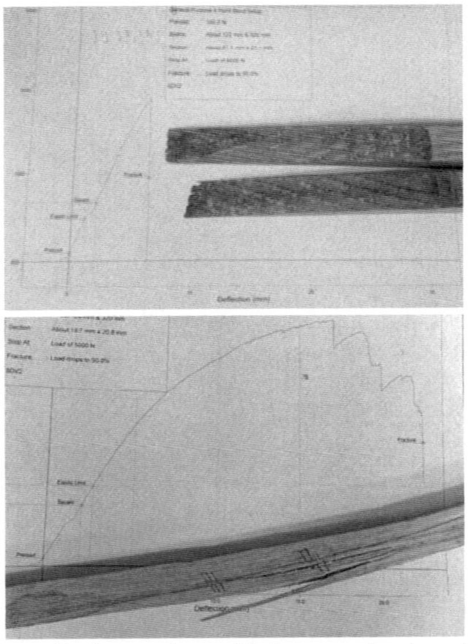

Figure 5: *Comparison of failure mechanisms [4].*

Further inspection of the samples revealed the presence of micro wood failures, specifically localized on the annual rings, as shown in Figure 6. It is hypothesized that the higher density of annual rings made for a lower resin absorption and therefore better bond.

Figure 6: *Localized micro wood failure [4].*

2.4 QUANTITATIVE RESULTS

Amongst the many mechanical properties that can be ascertained from this experiment, the two of primary importance here are the modulus of elasticity and ultimate flexural stress, respectively labelled as $E_{//}$ and $\sigma_{uf//}$, where the subscript '//' specifies that those properties are given parallel to the grain.

In the absence of mechanical testing, default values would be provided by the relevant rules and regulations. For small craft scantlings, Annex F of the ISO 12215–5 [13], specifies the default mechanical properties of typical wood species. Despite the recent revision of the standard [14], with updates on composites [15; 16] and commercial crafts [17], only minor changes were made to the default properties, and no modifications to the theory underpinning atypical species were implemented. Ultimately, regulatory bodies do not account for the presence of scarf joints or their ratios.

For a strength-driven design, where the primary concern is to ensure stresses remain below an acceptable level, the ultimate flexural strength will be utilized. Note that, in this instance, a safety factor would be employed to ensure added reliability, and that the material does not suffer from permanent deformation under normal loading. As a minimum requirement, the ISO 12215–5 imposes a factor of safety of 2 on the ultimate flexural strength, which leads to the design stress value, ultimately employed is the calculation of thickness calculation for panels, and section modulus for stiffeners. On the other hand, the modulus of elasticity comes into play for stiffness-driven designs, where the primary intent is to limit deflection to a comfortable level.

In structural testing, the final values for the mechanical properties are typically the lesser of either 90% of the average across all samples, or the average value achieved to which two standard deviations are subtracted, thus accounting for the scatter in the data. In all cases, the average minus two standard deviation proved to be the most pessimistic case, and thus was retained. Table 1 presents the average variation in two principal quantities of interest here, and demonstrates the conclusiveness of the results obtained.

Table 1: *Variance in the quantitative results.*

Samples	$\sigma_{uf//}$ **(MPa)**	$E_{//}$ **(MPa)**
Solid	7.88%	3.07%
4:1	8.66%	7.44%
8:1	7.07%	7.22%
12:1	5.74%	8.02%
16:1	1.70%	2.51%
20:1	3.73%	3.96%

The flexural strength and modulus of elasticity can then be plotted against the increasing scarf ratios of the samples, and compared to standard values given by structural regulations. Here, both the International Organisation for Standardization (ISO) [13] and the American Bureau of Shipping (ABS) [18] are considered, together with solid timber and published rule of thumb [8; 9].

The results are presented in Figure 7 as scarf efficiency, where 100% represents the strengths of the solid timber as determined experimentally, for the ultimate flexural strength and modulus of elasticity.

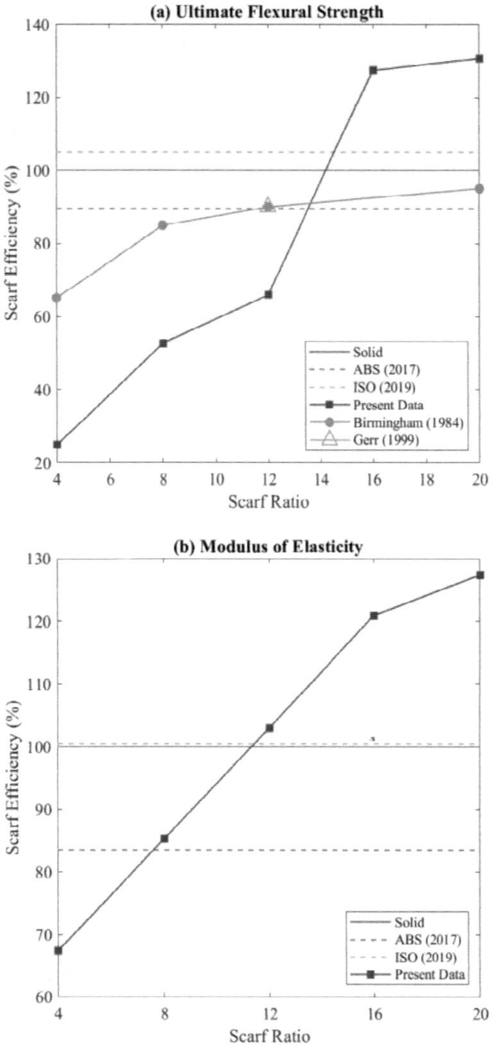

Figure 7: *Scarf efficiency compared to solid timber and typical regulatory values.*

The results reveal striking differences between the experimental data and both solid timber and the previously existing guidance. This implies that greater care should be taken in the design of structural components with a low scarf ratio, and thus a higher factor of safety should be used. Loss of strength was also noticed for small (6:1) Iroko samples [19], further confirming the present study. Conversely, the higher end of scarf ratios showed a significant improvement in mechanical properties, which could therefore be strategically utilized, particularly for weight-critical components, thus justifying the requirement for high ratios on spars.

With respect to the default properties provided by rules and regulations, that can be found in Table 2, there is a large difference in the actual values, with ABS [18] specifying more pessimistic mechanical properties and imposing a larger factor of safety than the ISO standard [13]. The latter also proved to be more in line with the results for solid timber.

Table 2: *Comparison of the experimental mechanical properties with solid timber and typical regulatory values.*

	Mechanical Properties	$\sigma_{uf//}$ (MPa)	$E_{//}$ (MPa)
	ABS (2017)	66.00	10 000
	ISO (2019)	77.00	12 060
Experimental	Solid Timber	73.51	11 980
	4:1 Scarf	18.28	8 045
	8:1 Scarf	38.86	10 270
	12:1 Scarf	48.72	12 379
	16:1 Scarf	93.43	14 486
	20:1 Scarf	96.11	15 250

2.5 CONCLUSIONS

When looking at an actual design, these research findings should be kept in mind. The examples here demonstrate that, in considering either ABS or ISO default ultimate flexural strength and inherent factor of safety, the design stress would have been over-estimated and thus would not have prevented failure of the 4:1 scarf ratio. Structural testing is a time-consuming and expensive approach; it is therefore hoped the results provided in this paper offer an efficient alternative and will allow designers and builders to adjust safety margins where necessary or help justify the need for an increased scarf ratio. This is particularly pertinent when tackling scantling determination for wooden boats [20].

The fact that these findings are specific to the timber species and adhesive tested here should be reiterated, and while qualitative similarities can be expected, quantitative results will require further research. Furthermore, it is vital to point out that the mechanical properties of wood can vary greatly and be affected by a wide range of parameters, including density, moisture content, grain orientation and straightness, defects, and so on, eventually leading to higher factors of safety when designing wooden boats.

The factor of safety adopted is also influenced by the thickness of wood: large sections carry greater uncertainty as to grain orientation and the presence of defects, which means they generally require an increased factor of safety. On the other hand, it is easy to spot any defect in thin pieces of wood. As a result, laminated components made of thin veneers generally permit reduced safety margins, allowing lighter and stronger structures. Consequently, laminated timber is a very attractive technique for modern construction, therefore calling for further experimental research in this field.

3. LAMINATED WOOD

3.1 BACKGROUND

The primary aim of this experiment was to characterize the mechanical properties of three species of timber present in Costa Rica, in order to support the current build of a wooden cargo sailing vessel, as well as providing the necessary data for new designs [21]. The three species under investigation are:

- *Cedrela Odorata*, $\rho \approx 548$ kg.m^{-3}
- *Cordia Gerascanthus*, $\rho \approx 661$ kg.m^{-3}
- *Dialium Guianense*, $\rho \approx 987$ kg.m^{-3}

In addition to solid samples, various adhesives will be employed for the laminated ones, namely:

- Epoxy (Ampreg 22)
- Resorcinol (Dynea Prefere 4050)
- Polyurethane (Geocel Joiner's Mater)

This will be compared to the values for solid timber. In the case of epoxy, two test batches will also be investigated, a standard one glued using clamps, and a more advanced manufacturing method, namely vacuum bagging.

From a regulatory perspective, both ISO [13] and ABS [18] assume greater overall properties for laminated timber. The former considering 50% of a timber's ultimate flexural strength when laminated (40% for solid), while the latter employs 42% of the modulus of rupture when laminated (37.5% for solid). This however does not account for the number of plies, adhesive or manufacturing method used. Hence the interest in performing destructive testing. Furthermore, previous work highlighted the need to treat laminated timber as composite laminates [22], contrary to the current regulatory process.

3.2 EXPERIMENTAL TESTING

3.2 (a) Manufacturing

All sample were manufactured from timber directly supplied by the Costa Rican shipyard to a final size of 400 mm long, by 20 mm wide, by 20 mm thick. In the case of the laminated samples, 5 layers of 4 mm were employed, glued with either epoxy, resorcinol or polyurethane, and clamped for the duration of the curing process. To replicate a more advance manufacturing process, laminated samples were also glued using epoxy under vacuum, as depicted in Figure 8.

Figure 8: *Laminated samples under vacuum bagging consolidation [23].*

3.2 (b) Method

The experimental campaign was undertaken under the specifications of the BS EN 408 standard [12], with the notable exception of a reduced number of samples. Indeed, due to the restricted supply of timber, only 4 samples were tested for each combination of the timber, glue and manufacturing method, thereby falling just short of the minimum 5 samples required. The four-point bending test employed on a Lloyds Instruments LR 30k tensile machine is depicted in Figure 9.

Figure 9: *Experimental setup [23].*

3.3 QUALITATIVE RESULTS

A comparison of the load-deflection curves for all three species, either solid timber or clamped epoxy laminated, are presented in Figure 10. The difference in behaviour between spices can immediately be identified, and is closely related to their respective density. Indeed, *Cedrela Odorata* is marginally less dense than *Cordia Gerascanthus*, with *Dialium Guianense* being far denser. In the case of *Cedrela Odorata* and *Cordia Gerascanthus*, the laminated sampled proved to reach failure at much lower deflections, though at virtually identical load for the latter. *Cordia Gerascanthus* also proved able to withstand a much higher level of deformation prior to rupture. Lastly, while comparable loads could be reached in the case of *Dialium Guianense*, the laminated samples proved more flexible, thereby allowing for greater deformation.

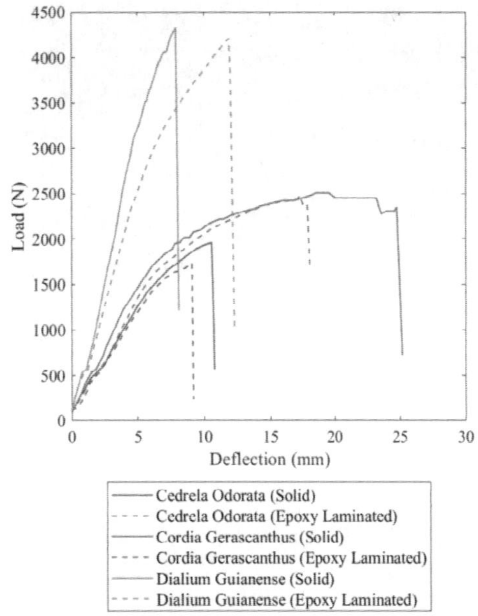

Figure 10: *Typical load-deflection curves [23].*

3.4 QUANTITATIVE RESULTS

The experimental data gathered allowed to characterise the two fundamental mechanical properties for strength and stiffness design, namely the ultimate flexural stress and modulus of elasticity respectively. As per Section 2.4, the final retained values are the lesser of either 90% of the average across all samples, or the average value achieved to which two standard deviations are subtracted.

In addition, the estimation of mechanical properties for rarer timber species provided in the ISO 12215-5:2019 [13] was implemented. Indeed, for unconventional species, the mechanical properties can be derived as a direct function of the density of the wood. Hence, the ultimate flexural strength ($\sigma_{uf//}$) and modulus of elasticity ($E_{//}$) of a hardwood of density ρ can be approximated as:

$$\sigma_{uf//} = 0.137\rho \tag{1}$$

$$E_{//} = 19.5\rho \tag{2}$$

The equations are slightly adjusted for softwood, and respectively given as:

$$\sigma_{uf//} = 0.130\rho \tag{3}$$

$$E_{//} = 17.5\rho \tag{4}$$

These estimates should however be treated very carefully, and mechanical testing should always be conducted to ensure the most suitable properties are employed as part of the structural design. The importance of this is demonstrated in the results for all three timber species depicted in Figure 11, with the numerical values summarised in Table 3, revealing extremely significant divergence in the actual and estimated properties for solid timber. The effect of the various adhesives and associated manufacturing techniques can also be observed.

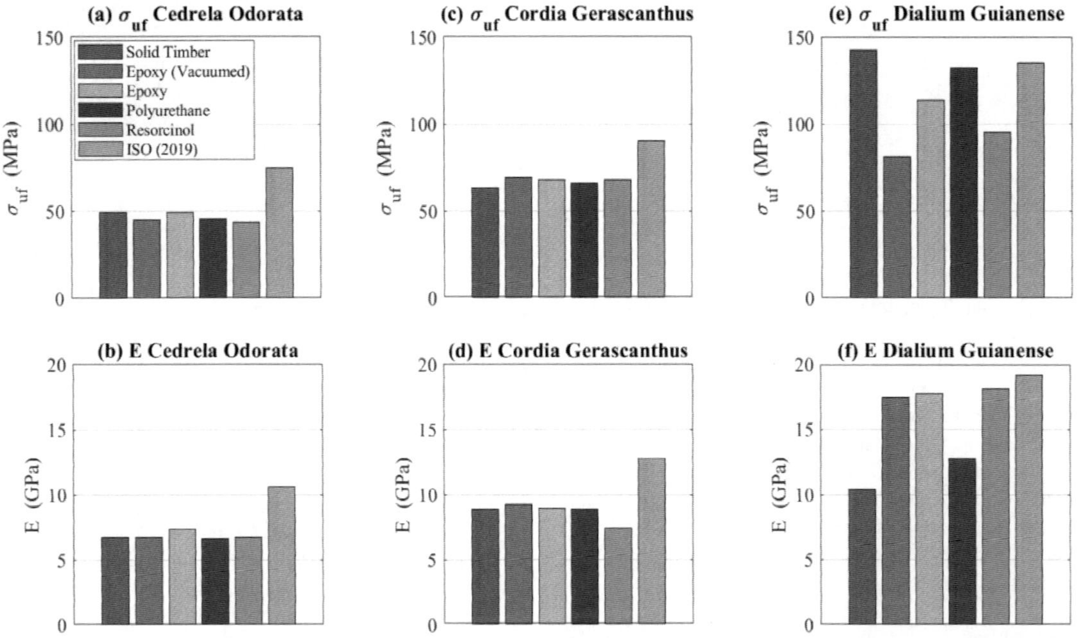

Figure 11: *Comparison of the ultimate flexural strengths and moduli of elasticity.*

Table 3: *Comparison of the experimental mechanical properties with solid timber and regulatory values.*

		Mechanical Properties	$\sigma_{uf//}$ (MPa)	$E_{//}$ (MPa)
Cedrela Odorata		**ISO (2019)**	75.01	10 686
Cedrela Odorata	Experimental	Solid Timber	48.20	8 510
Cedrela Odorata	Experimental	Epoxy (Vac.)	44.66	8 523
Cedrela Odorata	Experimental	Epoxy	48.68	9 340
Cedrela Odorata	Experimental	Polyurethane	44.58	8 474
Cedrela Odorata	Experimental	Resorcinol	42.93	8535
Cordia Gerascanthus		**ISO (2019)**	90.56	12 889
Cordia Gerascanthus	Experimental	Solid Timber	62.16	11 135
Cordia Gerascanthus	Experimental	Epoxy (Vac.)	67.97	11 579
Cordia Gerascanthus	Experimental	Epoxy	66.67	11 370
Cordia Gerascanthus	Experimental	Polyurethane	64.61	11 316
Cordia Gerascanthus	Experimental	Resorcinol	66.96	12 414
Dialium Guianense		**ISO (2019)**	135.22	19 246
Dialium Guianense	Experimental	Solid Timber	107.48	19 695
Dialium Guianense	Experimental	Epoxy (Vac.)	41.18	24 532
Dialium Guianense	Experimental	Epoxy	85.12	24 837
Dialium Guianense	Experimental	Polyurethane	100.11	21333
Dialium Guianense	Experimental	Resorcinol	71.88	25 085

3.5 CONCLUSIONS

In terms of the ultimate flexural stress, the results demonstrate that improvements can be achieved with laminated timber in the case of *Cordia Gerascanthus* (Figure 11 (c) and (d)), with further enhancement thanks to the vacuum bagging. These results are however very specific to each species, with *Cedrela Odorata* (Figure 11 (a) and (b)) displaying a loss of strength as a result on the lamination process (with the notable exception of clamped epoxy). Further and more significant loss of strength was noticed for *Dialium Guianense* (Figure 11 (e) and (f)), and is attributed to the extreme density of timber, that does not allow for suitable adhesive penetration and bonding.

The variations in stiffeness for both *Cordia Gerascanthus* and *Cedrela Odorata* remained minimal. However, strong improvements were revealed for *Dialium Guianense*, where the lamination process greatly enhanced the modulus of elasticity.

The present results clearly highlights that no generalisation can be made regarding the use of laminated timber, or adhesive type, as there is a crucial dependency on the actual timber species considered. In addition, care should be taken when looking at approximations for regulatory properties of unconventional timber species, as these proved too optimistic, and therefore unsafe, in all tested cases.

4. CONCLUSIONS

Timber construction remains strongly rooted in historical developments, with many traditional features still present in modern construction. Nevertheless, as more advanced wooden boats are designed and built, in line with relevant rules, it is critical to appraise the reliability of such regulations. This is particularly vital considering the complex nature of wooden designs, and comparatively lesser research undertaken compared to composites.

This paper presents the results of two experimental campaigns, the first focussed on the effect of scarf ratios, and the second tackling the effect of various adhesives for laminated unconventional timber species. The results show stark disparities with small craft regulations, and highlight the importance of undertaking destructive testing to characterize the mechanical properties, eventually feeding into the structural design process. Furthermore, it should be noted that extrapolation to other timber species did not prove straight forward, and thus care should be taken when dealing with different ones, particularly unconventional ones.

Ultimately, destructive testing would be strongly advised to support the structural design of wooden boats. Should this not prove feasible, additional factor of safety compared to that of regulatory bodies would be advised, as this paper demonstrated a number of limitations, where regulatory properties would appear far greater than the tested ones.

5. ACKNOWLEDGEMENTS

The significant contributions, from manufacturing to execution of the structural testing and post-processing by Stefan Scekic [3] and Ville Linden [23], for the scarfs and laminated timber respectively, is greatly acknowledged, and very much appreciated.

6. REFERENCES

[1] KWON, Y. W., 'Computational and experimental study of composite scarf bonded joints', *Composites Science and Engineering, Structural Integrity and Durability of Advanced Composites*, pp. 659-693, 2015.

[2] ISSC V.8 Committee, 'Sailing Yacht Design', *17th International Ship and Offshore Structures Congress*, Seoul, Korea, 2009.

[3] SCEKIC, S., 'Design of a 23 m modern-classic wooden sailing yacht with timber investigation', *Master Thesis, University of Liege*, 2018.

[4] SOUPPEZ, J.-B. R. G., 'Design and production of a one-off 32 feet wooden racing yacht dedicated to Southampton Solent University', *Southampton Solent University*, 2013.

[5] SOUPPEZ, J.-B. R. G., 'Design and Production of a Wooden Thames A-Rater Class Sailing Yacht', *Master Thesis*, 2015.

[6] KAROLAK, A., JASIENKO, J, NOWAK, T. & RASZCZUK, K, 'Experimental Investigations of Timber Beams with Stop-Splayed Scarf Carpentry Joints', *Materials*, vol. 13, 2020.

[7] LLOYD'S REGISTER OF SHIPPING, 'Rules and Regulations for the Classification of Yachts and Small Craft, Part 2, Hull Construction', Lloyd's Register of Shipping, 1979.

[8] BIRMINGHAM, R., 'Boatbuilding Techniques', 3rd edition, *Adlard Coles Nautical*, 2005.

[9] GERR, D., 'The Elements of Boat Strength', *International Marine, McGraw-Hill*, 1999.

[10] GOUGEON, M., 'The Gougeon Brothers on boat construction', 5th ed, *Gougeon Brothers*, 2005.

[11] CHAPELLE, H. I., 'Boatbuilding', *W. W. Norton & Company*, 1994.

[12] BRITISH STANDARDS, 'BS EN 408 - Timber structures - Structural timber and glued laminated timber - Determination of some physical and mechanical properties', *British Standards*, 2010.

[13] INTERNATIONAL ORGANISATION FOR STANDARDIZATION, 'BS EN ISO 12215-5:2019 - Small Craft – Hull Construction and Scantlings. Part 5: Design Pressures for Monohulls, Design Stresses, Scantlings Determination', *International Organisation for Standardization*, 2019.

[14] SOUPPEZ, J.-B. R. G. & RIDLEY, J., 'The revisions of the BS EN ISO 12215', *Marine Sector Showcase* - Composite UK, Southampton, UK, 2017.

[15] SOUPPEZ, J.-B. R. G., 'Structural Design of High Performance Composite Sailing Yachts under the New BS EN ISO 12215-5', *Journal of Sailing Technology*, issue 2, pp 1-18, 2018.

[16] SOUPPEZ, J.-B. R. G., 'Designing the Next Generation of Small Pleasure and Commercial Powerboats with the Latest ISO 12215-5 for Hull Construction and Scantlings', *1st SNAME / IBEX Symposium*, Tampa, Florida, US, 2019.

[17] SOUPPEZ, J.-B. R. G., 'Structural Analysis of Composite Search and Rescue Vessels under the New BS EN ISO 12215-5'. *SURV9 - Surveillance, Search and Rescue Craft*, The Royal Institution of Naval Architects, London, UK, pp 1-4, 2018.

[18] AMERICAN BUREAU OF SHIPPING, 'Guide for Building and Classing Yachts – Part 3 Hull Construction and Equipment', *American Bureau of Shipping*, 2017.

[19] BUCCI, V., CORIGALIANO, P., EPASTO, G. GUGLIELMINO, E. & MARINO, A., 'Experimental Investigation on Iroko Wood used in Shipbuilding', *Proceedings of the Institution of Mechanical Engineers, Part C, Journal of Mechanical Engineering Science*, 2017.

[20] SOUPPEZ, J.-B. R. G., 'Structural Assessment and Scantlings of Traditional Small Crafts', *Historic Ships, The Royal Institution of Naval Architects*, London, UK, 2020.

[21] LINDEN, V. & SOUPPEZ, J.-B. R. G., 'Sailing towards Sustainable Trading with Wooden Cargo Schooner', In the proceedings of the *British Conference of Undergraduate Research*, Sheffield, 2018.

[22] LOSCOMBE, R., 'Structural Design Considerations for Laminated Wood Yachts', *International Conference on the Modern Yacht*, Portsmouth, UK, 1998.

[23] LINDEN, V., 'Wooden Cargo Schooner for Sustainable Trading Initiatives', *Final Year Dissertation, Solent University*, 2018.

7. AUTHORS BIOGRAPHY

Jean-Baptiste R. G. Souppez is the Senior Teaching Fellow in Mechanical Engineering and Design at Aston University. As the UK Principal Expert in Small Craft Structures, he is in charge of representing the interest of the British marine industry in the development of international structural regulations (ISO12215) while also overseeing all small craft regulatory developments on behalf of the Institute of Marine Science, Engineering and Technology (IMarEST). He is also the Deputy Editor-in-Chief of the Society of Naval Architects & Marine Engineers (SNAME) Journal of Sailing Technology. His research in fluid dynamics features a wide range of experimental methods supporting his cutting-edge expertise in hydrofoils and award-winning work on the aerodynamics of downwind racing yachts sails. He contributes to the international Erasmus Mundus Master in Advanced Design in Ship and Offshore Structures (EMship+) as a Visiting Professor and Research Supervisor at the University of Liege, while promoting professional development as an Associate Tutor for the MLA College, supervising post-graduate researchers and enabling established professionals to achieve Chartered Engineer (CEng) status through the IMarEST.

The Royal Institution of Naval Architects
8-9 Northumberland Street
London, WC2N 5DA
United Kingdom

ISBN 978-1-7138-8299-2

CURRAN ASSOCIATES INC.
proceedings.com